Living Among Giants

Exploring and Settling the Outer Solar System

Living Among Giants

Exploring and Settling the Outer Solar System

Written and illustrated by Michael Carroll

 Springer

Michael Carroll
Littleton, CO, USA

ISBN 978-3-319-10673-1   ISBN 978-3-319-10674-8 (eBook)
DOI 10.1007/978-3-319-10674-8
Springer Cham Heidelberg New York Dordrecht London

Library of Congress Control Number: 2014950506

Printed on acid-free paper

Springer is part of Springer Science+Business Media (www.springer.com)

*This book is dedicated to the memory
of Lucien Rudaux, who first showed us what
things might be like out there,
and to Chesley Bonestell, who inhabited
those places with real people like you
and me.*

"Without adventure, civilization is in full decay."

– Alfred North Whitehead

Preface

Fig. P.1 Mars is the next logical site for human habitation. But what other sites offer promise? (Paintings ©Michael Carroll)

The outer Solar System may seem an unlikely destination for future human travels, let alone settlements. The cold and distant worlds of gas, rock, and ice seem to repel rather than to beckon. Humans have been to the Moon, and we have our sights firmly upon Mars and the asteroids, but could there be a role to play by the outer Solar System in the drama of humanity's future?

If humankind is to explore and eventually settle the outermost reaches of our Solar System, we cannot do it by bringing everything with us. The distances are far too great. We must, in a sense, live off the land. There is another destination, much closer, that offers us resources to do just that – the Red Planet, Mars.

A lot has been made of Mars. Although the Red Planet is not as close to Earth as Venus is, its atmosphere is clear enough to view the surface through telescopes. Mars intrigued at the outset, with its Earthlike seasonal tilt and day, its polar caps, and its mysterious, undulating dark regions. Flagstaff astronomer Percival Lowell brought the popularity of Martian canals to a frenzy, crafting detailed maps of Martian "canal" networks and "oases" and writing popular speculative books about life on Mars. Writers such as Wells, Burroughs, and Bradbury took their cue from the work of early astronomers such as Lowell,[1] continuing the nineteenth- and twentieth-century culture's love affair with the Red Planet.

We are left with the legacy of these writers and scientists. Most of the Mars they described has crumbled before the scrutiny of space probes, but our desire to find Martian life – even in microbial form – continues to inform, some would say contaminate, our priorities for space exploration. During a recent meeting of the American Astronomical Society's Division for Planetary Sciences, one researcher quipped, "If you compare what we know about Mars to what we know about Ganymede, it's shameful."

Still, if humans are to venture beyond the Moon for any length of time, the best place to go is one where we can live off the land. Mars has the resources to do so. It possesses large amounts of water, and its rarified atmosphere is 95 % carbon dioxide (CO_2), a molecular combination of carbon and oxygen. Water can be electrolyzed to produce hydrogen and oxygen. The hydrogen can be reacted with carbon dioxide from Martian air, to make methane, an efficient rocket fuel, and more water, which can be electrolyzed again. So the final products are methane and oxygen. Oxygen is something that most explorers prefer to breathe and is a major part of rocket propellant. Martian water is also useful for drinking and for tending plants, which themselves manufacture oxygen using the Martian CO_2.

Groups ranging from major aerospace corporations to private industry have carried out engineering studies in many forms, fleshing out scenarios for human Mars missions and settlements. They foresee the manufacture of Martian concrete and bricks to construct everything from rover garages to underground housing. Their crystal balls glimmer with visions of greenhouses full of bamboo, which grows fast and makes strong

1. Lowell also helped to discover the dwarf planet Pluto. Using observations of Uranus and Neptune, he predicted where a new planet should be and searched in vain for it. However, it was later discovered that Lowell Observatory captured the planet in photos in 1915, a year before Lowell's death. Pluto was discovered by Clyde Tombaugh in 1930. In recognition of Lowell's contributions to its search, the planet's symbol incorporates the initials of Percival Lowell, PL.

building material. Some even envisage genetically engineered goats, tilapia, strawberries, and poi to feed the masses of incoming Earth settlers.

Mars seems the next logical step. But Mars is taking the lion's share of planetary probe pie, and many believe it is time to shift a few more of our resources farther out. "What about Venus?" some ask. After all, the hellish world is Earth's twin in size, closest to us in distance, and complex both geologically and atmospherically. But Venus will be no home for humans in the foreseeable future using reasonable technology. All of our landers have succumbed to its 900 °F surface temperatures in just over an hour, and even the most current advances in engineering could not extend that stay much beyond a day or two. Brutal pressures claw at electronics, and acids eat away at insulation, while the heat demands cooling energy that might be used otherwise. It is not a nice place to visit.

However, if we look the other direction, beyond Mars, we find surprisingly rich resources and promise. True, the outer Solar System is dark and bitterly cold. Distances make communication and travel difficult. But the outer worlds, giants of gas and ice, possess entourages of icy and rocky moons replete with water, minerals, and hydrocarbons, and it is there where we may find a new future.

Arthur C. Clarke believed that the realm of the outer planets was a natural destination for humankind. He said, "We are exiles here on dry land, in transit between the ocean of water in which we were born and the ocean of space where most of history will run its course."

Mars settlement advocate Robert Zubrin takes it a step further, asserting that we *must* go. "The Hawaiian islands popped out of the ocean. The birds flew overhead and dropped seeds, and brought life to those places. There is oxygen in the air because life put it there. There is soil on the ground because life put it there. This is what we do. It would be unnatural if humans, being the kind of bird that the biosphere has developed to spread life across space, didn't drop the seeds of life on the desert islands out there in the cosmos."

If people like Clarke and Zubrin are right, Mars will be a stepping-stone along the way to farther shores. The worlds awaiting us beyond, in the frigid darkness, harbor abundant natural resources and deep mysteries that will likely lead to our foundational understanding of our planetary system. Carolyn Porco, Principal Investigator for the Cassini Saturn Orbiter imaging team, puts it this way: "No matter how you measure it, whether you count the number of bodies, whether you add up the amount of mass, or whether you calculate the volume taken up by the orbits of those bodies, the vast majority of our Solar System lies out beyond the orbit of the asteroids. Inside are just a bit of flotsam. It's all in the outer Solar System."

Moreover, future travelers to the realm of the gas and ice giants will be confronted by glorious, spectacular views beyond anything experienced thus far. Aside from practical scientific and technological gain, travel to the icy cliffs, thundering geysers, incandescent volcanoes, and swirling storms will bring inspiration. As artist/explorer Frederick Church once said,

"Exploration is good for the soul." Henry David Thoreau advised that: "We need the tonic of wildness … At the same time that we are earnest to explore and learn all things, we require that all things be mysterious and unexplorable, that land and sea be indefinitely wild, unsurveyed and unfathomed by us.…" Although we've plastered our Earth maps to their corners with information from satellite and ground surveys, our maps of the worlds beyond still have vast territories labeled *Thar be dragons*. The cosmos compels us to fill them in. We have gone there with our robots, but the history of space exploration shows us that where our robots go, the footprints of humans will follow.

Littleton, CO, USA   Michael Carroll

About the Author

Springer author/artist Michael Carroll received the AAS Division of Planetary Science's Jonathan Eberhart Award for the best planetary science feature article of 2012, an article based on his Springer book *Drifting on Alien Winds*. He lectures extensively in concert with his various books and has done invited talks at science museums, aerospace facilities, and NASA centers. His two decades as a science journalist have left him well connected in the planetary science community. He is a Fellow of the International Association for the Astronomical Arts and has written articles and books on topics ranging from space to archaeology. His articles have appeared in *Popular Science*, *Astronomy*, *Sky & Telescope*, *Astronomy Now* (UK), and a host of children's magazines. His 20-some titles also include *Alien Volcanoes* (Johns Hopkins University Press), *Space Art* (Watson Guptill), *The Seventh Landing* (Springer 2009), and *Drifting on Alien Winds* (Springer 2011). His latest coauthored book is Springer's *Alien Seas: Oceans in Space* (2013).

Carroll has done commissioned artwork for NASA, the Jet Propulsion Laboratory, and several hundred magazines throughout the world, including *National Geographic*, *Time*, *Smithsonian*, *Astronomy*, and others. One of his paintings is on the surface of Mars – in digital form – aboard the Phoenix lander. Carroll is the 2006 recipient of the Lucien Rudaux Award for lifetime achievement in the astronomical arts.

A Note About the Paintings

Living Among the Giants has a dozen or so original paintings done specifically for the project. Almost all are traditionally done on canvas. We live in a digital world, and this book is about technologically advanced things, so I decided that traditional paintings would bring a visual softness to the subject. Some are "tradigital," such as the painting of the astronaut on Titan. In this case, I began with a traditional painting on board and then added digital touches in Photoshop and Terragen. I hope my readers will enjoy the results!

Acknowledgments

My artistic thanks goes to Carolyn Porco of the Space Science Institute and Joe Spitale of the Planetary Science Institute for their patience and wisdom on Saturn's ring "spikes." A special shoutout goes to Rob Callison for ideas on submarine psychology. For translations of Lucien Rudaux's work, I am indebted to Jenna Khazoyan and Caroline Carroll. Bill Higgins was of tremendous help in providing material and insights into the history of science fiction. Thanks to Wes Patterson and Chris Paranicas at Johns Hopkins Applied Physics Laboratory for use of their Europa sputtering diagram (if you don't know what that is, you'll find out in Chap. 5). Ted Stryk generously lent his digital magic to images of various planets and moons. Thanks to Keith Cooper and the team at *Astronomy Now*, and to Dave Eicher and the gang at *Astronomy*, for letting me borrow from articles written for them. Cynthia Rodriguez and Princess Cruise Lines generously gave permission for me to use their logo and beautiful ships as reference for our own cosmic cruise ship in Chap. 10. Aldo Spadoni brought life to the Iapetus Ridge Resort, aka "New Santorini," and Edie Carroll (my very own Mom), Marilyn Flynn (special commendation for last-minute frenzy), and Bill Gerrish made my words look intelligent, more or less (there's only so much one can do). My talented daughter Alexandra helped transcribe interviews, a critical part of any book like this. Thanks to Alice Salvage, Chris White, and Matt Levin at Magnolia Pictures and Wayfare Entertainment for the cool scene from *Europa Report*.

And many thanks to Maury Solomon – fearless editor at Springer – for her years of support.

Contents

Part I
The Backdrop

Fig. 1.1 *The Flemish painter Garrit Dou (1613–1675) crafted this masterpiece of an astronomer working by candlelight. The scholar's tools include a liquid-filled beaker, an hourglass, a huge book, and a celestial globe showing the constellations of the night sky. Although telescopes were invented at the opening of the seventeenth century, most early astronomers did not have access to them; the objects in the painting were the tools of their trade. At the end of the day, the ancient student of the sky could only dream of what travelers might discover out there (Painting by Garrit Dou, ca. 1658)*

Chapter 1
Early Ideas

In the first century A.D., Greek Hellenistic thought and culture continued a centuries-long spread throughout the western world. While the Mayan kingdom debuted its long count, China's Han dynasty officially adopted Confucianism, and a Jewish rabbi was stirring things up in the Middle East, Lucian of Samosata was quietly writing a story about travel to the Moon. As with much fiction, his was more a commentary and satire of contemporary literature than a narrative of imagined events. Lucian didn't have much hard science to go on in the first century.

Lucian's *True History* detailed a nautical voyage of discovery interrupted by a violent waterspout. After a harrowing, windy week, the narrator and his intrepid crew found themselves deposited on the Moon. Lucian described his voyage:

> [O]n the eighth day we saw a great country in it, resembling an island, bright and round and shining with a great light. Running in there and anchoring, we went ashore, and on investigating found that the land was inhabited and cultivated. By day nothing was in sight from the place, but as night came on we began to see many other islands hard by, some larger, some smaller, and they were like fire in color. We also saw another country below, with cities in it and rivers and seas and forests and mountains. This we inferred to be our own world.[1]

Lucian's story, written some nineteen centuries ago, includes elements of space travel, encounters with aliens, artificial atmosphere, and even the scientific desire to explore and discover. The Moon continued to be the central object of cosmic speculation for centuries to come.

In September of 1610, German astronomer and mathematician Johannes Kepler wrote a short paper confirming Galileo's discovery of Jupiter's four large moons. It was called, simply, *Observation-Report on Jupiter's Four Wandering Satellites*. The pamphlet marked the first time that anyone had referred to moons of other planets as "satellites." Kepler may have had in mind Jupiter's royal standing as king of the planets. In Kepler's day, heads of state and other important figures surrounded themselves with a cadre of fans doubling as bodyguards. These early paparazzi were known by the Latin word *satellitem*. (Kepler had used the term in a letter to Galileo as well.)

Kepler didn't limit his writing to scientific journals. Over a period of 20 years, he transformed his 1608 student dissertation into what many consider to be the first work of modern science fiction. Called *Somnium*, the story was actually published after Kepler's death in 1630. Kepler's narrative opened as he was reading a book by a magician. He fell asleep (the magician's writing must not have been very engaging) and dreamed of an Icelandic boy taken to the Moon by daemons, kindly spirit-guides. Inhabitants of the Moon called it Levania.

Kepler's story anticipated many scientific issues, such as atmosphere (his travelers must have moist sponges over their nostrils to breathe), the cold of space (travelers wrap themselves in blankets), and great distances

1. Internet Sacred Text Archive translation, public domain. http://www.sacred-texts.com

M. Carroll, *Living Among Giants: Exploring and Settling the Outer Solar System*, DOI 10.1007/978-3-319-10674-8_1, © Springer International Publishing Switzerland 2015

between Earth and the Moon. Like any good science fiction, Kepler's writing echoed scientific thinking of the time. It described the appearance of eclipses from the Moon; it portrayed the size of planets as changed because of the Moon's distance from Earth; it even gave a guess as to the size of the Moon and described the effects of lunar tidal lock (the phenomenon of the same side of the Moon always facing Earth). Sadly, Kepler awakened before we could learn about such subtleties as good landing sites and lunar mascons.

It would be many decades before writers tackled the nature of the outer Solar System, let alone the idea of exploring those distant worlds. Dutch mathematician Christian Huygens was undaunted by distances and rudimentary instruments. Huygens studied the cosmos at the end of his 50-power telescope, where he made such discoveries as Saturn's moon Titan, the length of a Martian day, and details of the Orion and other nebulae. In his 1698 book *The Celestial Worlds Discover'd: Or, Conjectures Concerning the Inhabitants, Plans and Productions of the Worlds in the Planets*, Huygens examined the nature of outer planets, especially as they related to clouds and water:

> [A]bout Jupiter are observ'd some spots of a darker hue than the rest of his Body, which by their continual change show themselves to be Clouds … Since 'tis certain that Earth and Jupiter have their Water and Clouds, there is no reason why the other Planets should be without them.… this Water of ours, in Jupiter or Saturn, would be frozen up instantly by reason of the vast distance of the Sun. Every Planet therefore must have its Waters of such a temper, as to be proportion'd to its heat.

Significantly, Huygens went on to speculate about the moons of the outer planets.

> [A]ll the Attendants of Jupiter and Saturn are of the same nature with our Moon, as going round them, and being carry'd with them round the Sun just as the Moon is with the Earth. Their Likeness reaches to other things, too … whatsoever we can with reason affirm or fancy of our Moon (and we may say little of it) must be suppos'd with very little alteration to belong to the Guards of Jupiter and Saturn, as having no reason to be at all inferior to that.

The idea that the moons of Jupiter and Saturn are similar to our own Moon carried well into the decade of the 1960s. In intervening years, literary characters visited our own Moon,[2] saw dinosaurs on Jupiter,[3] lived in colonies on Ganymede,[4] passed Saturn on a comet,[5] and witnessed the first human expedition to Neptune in Hugh Walter's 1968 novel *Nearly Neptune*.

Mars continued to be the driver of interplanetary travel tales, with Percival Lowell unwittingly at the helm. In his wildly popular and supposedly non-fiction book *Mars as the Abode of Life*, Lowell declared:

> Thus, not only do the observations we have scanned lead us to the conclusion that Mars at this moment is inhabited, but they land us at the further one that these denizens are of an order whose acquaintance was worth the making.[6]

2. For example, *From the Earth to the Moon* (1865) by Jules Verne.

3. *A Journey in Other Worlds* (1894) by John Jacob Astor IV.

4. *Victory Unintentional* (1942) by Isaac Asimov.

5. *Off on a Comet* (1877) by Jules Verne.

6. *Mars as the Abode of Life* by Percival Lowell (The MacMillan Company, 1909).

Fig: 4.

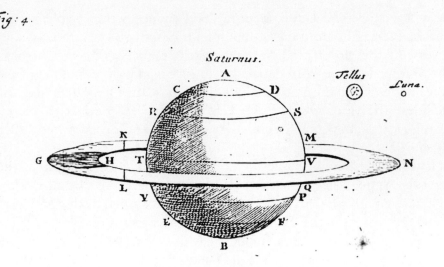

Fig. 1.2 *Drawing of Saturn and its rings by Christiaan Huygens, ca. 1659. Note his comparison to Earth at right (Image from Systema Saturnium)*

The novelist Edgar Rice Burroughs, famous for his Tarzan tales, fully embraced Lowell's thoughts on Mars, spinning a series of narratives populated by six-legged tiger-like Thoats, green four-armed Martian insect men, and of course, a beautiful princess, Dejah Thoris.[7]

Burroughs paints a picture of Dejah Thoris and her Martian civilization as attempting to preserve Mars' dying atmosphere, something Lowell's observations certainly inspired. Here, she berates the brutish green Martian race for their lack of understanding:

> The work we were doing was as much in your interests as in ours, for you know full well that were it not for our labors and the fruits of our scientific operations there would not be enough air or water on Mars to support a single human life. For ages we have maintained the air and water supply at practically the same point without an appreciable loss, and we have done this in the face of the brutal and ignorant interference of you green men. Why, oh, why will you not learn to live in amity with your fellows?

Burroughs wrote what some would call informed fantasy, but H. G. Wells was more interested in the science of his time. Where Burroughs' John Carter simply dreams his way to Mars, Wells' books outline the details of spacecraft and space travel. Martian anatomy is described in detail, tied to contemporary scientific assumptions about the Martian environment.

H. G. Wells' Martians "drew their plans against us," but 30 years later, as a real Cold War was ramping up on Earth, the Martians of Ray Bradbury became far more sedate, and certainly more mystical:

> They had a house of crystal pillars on the planet Mars by the edge of an empty sea … Mr. and Mrs. K had lived by the dead sea for twenty years, and their ancestors had lived in the same house, which turned and followed the sun, flower-like, for ten centuries. They had the fair, brownish skin of the true Martian, the yellow coin eyes, the soft musical voices.

Although the literature reflected scientific thinking of its time, the outer planets offered little fodder for hungry authors. The distant worlds had given up few secrets over the centuries. Still, writers dreamed of that

7. Originally published as a serial in the magazine *The All-Story*, beginning in February 1912.

Stygian outer realm. French enlightenment philosopher Voltaire (Francois-Marie Arouet) put his hand to writing a narrative called *Micromegas*, published in 1752. His is the story of an alien being from a distant star who travels through our Solar System by means of comets and sunbeams. The alien is 8 leagues tall (24 miles) and roams with a sidekick from Saturn, a diminutive little being merely a mile in stature. Voltaire then details the adventures of the two travelers: "[T]hey first of all jumped upon Saturn's ring, which they found pretty flat, as an illustrious inhabitant of our little globe has very cleverly conjectured … they came across the satellites of Jupiter. They landed on Jupiter itself…"

Voltaire had a clear grasp of contemporary science. His quip about the "illustrious inhabitant of our globe" refers to Christiaan Huygens, the astronomer who had, only decades earlier, published his discourses on Saturn's rings and discovered its largest moon Titan.

Three decades after *Micromegas*, an author writing under the pseudonym Vivenair penned a short story entitled "A Journey lately performed through the Air, in an Aerostatic Globe … To the newly discovered Planet, Georgium Sidus." Georgium Sidus was the unofficial name first given to Uranus shortly after its discovery.

In the next century, Jules Verne took his readers on flybys of Jupiter and Saturn in his book *Off on a Comet*. His was the first attempt at a scientifically accurate account of what humans might encounter in the outer Solar System. Some of the science seems quaint to us. For example, at the time it was thought that the distant outer planets were more ancient than the ones closest to the Sun. Verne explains to his readers that:

> Neptune, situated 2,746,271,000 miles from the sun, issued from the solar nebulosity, thousands of millions of centuries back … Jupiter, the colossal planet, gravitating at a distance of 475,693,000 miles, may be reckoned as 70,000,000 centuries old … and Mercury, nearest of all, and youngest of all, has been revolving at a distance of 35,393,000 miles for the space of 10,000,000 years – the same time as the moon has been evolved from the earth.

As Verne's travelers approach Jupiter, they begin to see details in the planet's clouds as well as subtleties across the surfaces of its moons:

> Jupiter began to wear an aspect that must have excited the admiration of the most ignorant or the most indifferent observer. Its salient points were illumined with novel and radiant tints, and the solar rays, reflected from its disc, glowed with a mingled softness and intensity. And what an increased interest began to be associated with the satellites! They were visible to the naked eye! Was it not a new record in the annals of science?
>
> …[H]ere, at least … everyone could so far distinguish them one from the other as to describe them by their colors. The first was of a dull white shade; the second was blue; the third was white and brilliant; the fourth was orange, at times approaching to a red. It was further observed that Jupiter itself was almost void of scintillation.

Verne's tale also enables the author to speculate about the experience of seeing Saturn's rings from the "surface" of the planet (which we can read as the cloudtops):

To any observer stationed on the planet, between the extremes of lat. 45 degrees on either side of the equator, these wonderful rings would present various strange phenomena. Sometimes they would appear as an illuminated arch, with the shadow of Saturn passing over it like the hour-hand over a dial; at other times they would be like a semi-aureole of light. Very often, too, for periods of several years, daily eclipses of the sun must occur through the interposition of this triple ring…. Truly, with the constant rising and setting of the satellites, some with bright discs at their full, others like silver crescents, in quadrature, as well as by the encircling rings, the aspect of the heavens from the surface of Saturn must be as impressive as it is gorgeous.

Jules Verne's account provided us with our first clear glimpse of human experience among the giants of our Solar System. But his mode of transportation left something to be desired.

At about the same time the writer Raoul Marquis, publishing under the pseudonym G. Le Faure and Henri de Graffigny, published a four-volume novel called *The Extraordinary Adventures of a Russian Scientist.*[8] The astronomer Camille Flammarion wrote the introduction. The books present a variety of novel ideas for spacecraft designs, including a craft made for a trip from Mars to Jupiter. The spaceship was barrel-shaped, 5 m across and 7 m long. A propeller inside pulls interplanetary material into a pipe and shoots it out the back, a design that anticipated jet engines and ramjet spacecraft power plants.

The pace picked up as readership became aware of the outer planets as real worlds where humans might explore and live. The year 1894 saw the publication of John Jacob Astor's novel *A Journey in Other Worlds.* His action took place in the year 2000, when antigravity spaceships made their way to Jupiter and Saturn. The 1930s saw pulp writer Stanley G. Weinbaum exploring the outer planets in his short story "Flight on Titan."[9] Weinbaum's work reflected the popular scientific theory of the time that the Solar System was born of a close encounter with another star, leaving the gas giants hot enough to provide Earth-like climates on Jupiter's moons Io and Europa, and on Saturn's moon Titan, and even on Uranus. In fact, Weinbaum wrote a book dedicated to the greenish planet in 1935. He called it *Planet of Doubt.*

In 1902, live action came to the tales of human space exploration with Georges Méliès' film *Le Voyage Dans La Lune* (A Trip to the Moon). The screenwriter's five astronomer-astrologers built a bullet-shaped capsule and climbed aboard. After their launch by cannon, they met up with insect-like Selenites, the King of the Moon, and Phoebe, goddess of the Moon, who seemed to sing to them. Sadly, movies with sound were still decades away.

Perhaps in an attempt to cash in on some of Melies' success, Spanish director Segundo de Chomon created the film *Voyage à Planète Jupiter.* His cosmic voyager – the king of the Earth – made his trip to Jupiter in a dream. Jupiter turned out to be a bizarre mix of a cloudy world and mythological home to the god who throws thunderbolts. After Verne's science, de Chomon's work seemed a step back into fantasy, but his was a great experiment in the science of filmmaking.

8. *Aventures extraordinaires d'un savant russe* by G. Le Faure and Henri de Graffigny, 1889.

9. *Astounding Magazine,* January 1935.

Fig. 1.3 Left and center: The first space travelers land – bull's eye! Scenes from Georges Méliès' 1902 film Le Voyage Dans La Lune. Right: The King of Earth passes Saturn on his way to Jupiter, climbing a cosmic ladder in de Chomon's dream film Voyage à Planète Jupiter

Fig. 1.4 Scene from the 2013 film Europa Report. Note the realistic treatment of the surface ice, Jupiter, and the suit technology, realism found consistently throughout the film (Image courtesy of Wayfare Entertainment and Magnet Releasing. Used with permission)

Hollywood had been in the act for decades, but at the time of Hugh Walter's 1968 novel, a new film depicted – for the first time – a highly accurate vision of a human expedition to Jupiter. The film was *2001: A Space Odyssey*, inspired by Arthur C. Clark's *The Sentinel*. The film depicted a Jupiter-bound vessel, propelled by nuclear-powered engines. The spacecraft design included an internal section that spins for artificial gravity and hibernation pods for some of the crew. Its sequel, *2010: Odyssey Two*, made use of data from the then-recent Voyager encounters of the Jupiter system. 1972's *Silent Running* portrayed a future when all forests of Earth reside in domed ships in the neighborhood of Saturn. The old-west-meets-space 1981 film *Outland* found its setting on Jupiter's moon Io, and contained some references to scientific knowledge of the time, although with liberties taken.

The most accurate film to date pertaining to outer planet exploration is Sebastian Cordero's gritty *Europa Report*. The film portrays a privately funded venture to search for life on Jupiter's oceanic ice moon. JPL scientists Steve Vance and Kevin Hand provided technical and scientific advice.

From *Somnium* to *Europa Report*, the media has come a long way in depicting outer planet exploration, and those depictions have influenced our cultural awareness of the vast Solar System beyond Mars. But that outer system is foundationally different from Mercury, Venus, Earth, the Moon and Mars.

THE TERRESTRIAL PLANETS

The inner planets are called terrestrials because of their similarities to Earth. Unlike the outer giant worlds, the terrestrials all have solid surfaces. And while the inner planets may differ fundamentally from the gas and ice giants, we will find some spooky echoes of them in the moons of the outer Solar System. Understanding the hot, rocky worlds will lend insight into understanding the outer worlds of rock and ice, our ultimate destinations.

The surface of Earth, stable and solid as it may seem, displays tatters and scars from a long succession of assaults. One of the most important of these assaults comes in the form of tectonics. Earth's crust is not of one piece. The continents rest upon plates that drift and pirouette about at the rate of the growth of a human fingernail.

Although many geologists remarked on the fact that the continents seem to have shifted through time, the real force behind the modern view of plate tectonics was the German scientist Alfred Wegener. While recovering from a neck wound sustained on the battlefields of World War I, Wegener came upon the idea that Africa and South America, along with all the other landmasses of the world, had once formed a supercontinent that he called *Pangaea* (*Pan* meaning "all," and *gaia* meaning "Earth"). He searched the world over for proof of his model. He matched the profiles of mountain ranges in South Africa and Argentina. He linked a plateau in Brazil with another on the Ivory Coast of Africa. He demonstrated geographic links of the fossil fern Glossopteris. Its locations line up perfectly with the coastlines of the two continents. Wegener wrote that, "It is just as if we were to refit the torn pieces of a newspaper by matching their edges and then check whether the lines of print run smoothly across … there is nothing left but to conclude that the pieces were in fact joined…"

Wegener's shifting continents were a bit like Copernicus' hurtling Earth; the idea went against the senses. Despite his proofs, there were skeptics. The idea, when first proposed, was met with widespread derision. The scientific community saw Wegener's evidence as largely circumstantial. One reviewer of his book on the subject said, "If we are to believe Wegener's hypothesis, we must forget everything which has been learned in the last 70 years and start over again." As it turns out, the critic was absolutely right. But it was the science of the past 70 years that had been mistaken.

By the 1960s, the scientific community was beginning to assemble a picture of the ocean floor. Long thought to be a lifeless desert of rolling plains and a few seamounts (underwater mountains), the sea floor was turning out to be bizarre and unlike conventional mountains and deserts. Sonar maps of the Mid-Atlantic Ridge revealed an undersea mountain chain unlike anything seen before. The long ridge rose 6,000–10,000 ft from the sea floor, and had a deep cleft or valley along its top. Unlike the mountain ranges of the continents, which seemed to have been pushed up by a shrinking Earth, this set of sinuous mountains had evidence of stretch all along it. Even more puzzling was the way these ridges wandered across Earth's ocean basins; at times they continued right up onto the land in some places such as Africa's Great Rift Valley and Thingvellir in Iceland. In others, the chain of submerged peaks echoed the coastlines of the Americas to its west and Africa to the east. How could the world be shrinking *and* stretching?

The answer lies in the constant shifting of Earth's crustal plates, which lead to a host of geologic forms. In some areas, the plates bash into each other, thrusting up mountain ranges like the Himalayas. In others, they rub along each other, causing faults, earthquakes and volcanoes. Still other areas spread apart (as at the Mid-Atlantic Ridge), picking up the slack for another kind of tectonics, subduction, where one plate dives under another.

Earth's brand of jigsaw crust seems unique in the Solar System, but aboveground all the terrestrial planets have something in common: they all have thin to no atmospheres. The densest of these is Venus, whose air pressure at the surface is similar to the water pressure on Earth's ocean floors. Earth has the next densest, with Mars a distant third (the Martian surface pressure is equivalent to Earth's at 100,000 ft altitude). The atmospheres of the inner worlds are dominated by carbon dioxide gas, with the odd mix of nitrogen and oxygen being unique to Earth. Mercury's vacuum matches that of our own Moon.

The young Sun's energetic solar winds blew the early atmospheres away, but the large enough planets – the ones with more gravity – held on

Fig. 1.5 Iceland's Thingvellir valley is an extension of the great rift traveling down the spine of the Mid-Atlantic Ridge. The two plates making up the valley walls are moving apart at the rate of 7 mm per year (Photo by the author) (This is an added piece of art not accounted for)

to their own air expelled through volcanoes and other processes. These new atmospheres would morph, through chemical and biological processes,[10] into the ones we see today.

All four planets have gone through an early phase in which their surfaces were covered with oceans, not of water but of molten rock. The Moon's dark "seas" that lovers (or werewolves) gaze upon each night were, at one time, incandescent oceans of lava erupted from volcanoes or exhumed by infalling meteors and comets as they broke through the rock crust. In fact, the craters left over from this violent era and ensuing time give researchers an idea of how geologically young a surface is.[11] The more craters, the less the surface has been transformed since its creation. Crater counts are an important tool for deciphering the past of a planet or moon.

Crater counts also tell us something about a planet's atmosphere. Atmospheres block incoming meteors and comets. Air friction vaporizes their surfaces, so that only the largest make it through. Venus has very few craters on its surface. Instead, starbursts of dark material paint areas on the surface where meteors exploded overhead. The few large craters that have survived are heavily eroded and transformed by Venusian volcanoes and wind. Mars has areas completely devoid of craters, and others where the surface appears pummeled and ancient. Its thin atmosphere cannot hold back all the incoming impactors, but its winds go to work to wipe them out over time.

Craters can be obliterated by many processes. Wegener's plates have destroyed countless impact features over the ages. Volcanoes erupt, blanketing whole regions with ash and lava. Glaciers (the result of long-term snowfall) and floods scour vast landscapes.

Earth's weather plays an important role in sculpting its surface. Rains and snows tear at solid rock. Aeolian (wind) erosion works its magic gradually, over long periods. Heat and cold cycles break down stone and gravel. Hurricanes change entire coastlines.

For its part, the weather on Venus appears to be far more gradual. Its soupy atmosphere moves in great waves, but winds at the surface are fairly gentle. Sulfuric acid and chlorine eat away at basaltic rock. Some areas may even exhibit metallic "frost" condensing on mountain slopes in the 900 °F climes.

Martian weather is extreme in its own way. Although surface wind gusts reached 27 m/s (60 mph) at the Viking sites, the Pathfinder lander found winds generally below 8 m/s (~18 mph). The thin air prevents Aeolian processes from taking a quick toll. Still, wind is the primary eroding force, since the Red Planet has probably not seen rainfall in billions of years. The only precipitation on Mars today is an occasional CO_2/H_2O snow flurry at its poles.

Some planets retain more of an atmosphere because of a magnetic bubble, or magnetosphere, which protects them from the relentless flow of solar particles. A magnetosphere is generated by the ebb and flow of a molten metal sea surrounding a solid core. The electrically conducting

10. At least in the case of Earth. Our oxygen/nitrogen atmosphere is the direct result of biological processes. We do not find free-floating oxygen on any other terrestrial planet.

11. In the Solar System's early days, mountain-sized leftovers regularly slammed into planetary surfaces, leaving vast impact scars. This heavy bombardment seems to have tailed off around 3.8 billion years ago, but cosmic debris continues to batter planets and moons at a steady rate. Just ask the Russian people. Two of the most famous meteoric interlopers streaked through their skies in modern times at Tunguska and Chelyabinsk, the latter caught on a host of dash cams. Orbiters such as ESA's Mars Express and NASA's Mars Global Surveyor and Mars Reconnaissance Orbiter have charted 200 impacts each year on the Martian surface. Even with Earth's dense atmosphere, anywhere from 5 to 300 t of meteors and cosmic dust make it to Earth's surface each day.

Fig. 1.6 Three systems for measuring Martian winds: Left: The Viking lander's meteorology boom. Center: Wind "socks" aboard the Pathfinder lander. Right: Wind sensor atop the Phoenix lander's meteorology boom (Images courtesy of NASA/JPL)

liquid metal sends out fields of energy much like a bar magnet. But not all planets have an electrified shield. This is due, in part, to each planet's development and size.

Early in our Solar System's history, the inner planets accreted, or fell together, from a cloud of dust and gas surrounding the infant Sun. As they grew, heavier materials sank to the core, while the lighter stuff formed a crust. During this formative epoch, internal heat from the condensation of the terrestrials combined with radioactive materials to warm the interiors enough that metals liquefied and made their way to the center. The larger the planet, the more radiogenic heating took place. So Earth and Venus had the most, our Moon the least; Mars and Mercury fell somewhere in between.

Radiogenic heating continues today on Earth and Venus, but it is not known if it plays an important part on Mars, which appears to be geologically quiescent. Mercury has a weak active magnetic field around it, which appears to be generated by molten metal on the inside, perhaps kept alive by continued radiation from materials within.

Spacecraft can detect the magnetic signatures of planets and moons as they fly past. Magnetic fields are a tricky indicator of internal activity, though, as they are affected by several factors. Earth has an active one, shielding us from the Sun's deadly UV radiation. Venus is identical to Earth in size but has no magnetic field. It certainly has enough radiogenic heating to preserve a molten core, so what happened? A liquid metal core must be stirred to generate a healthy magnetic field. While Earth and Mars both spin around in roughly 24 h, Venus takes its time, with a leisurely rotation of 243 Earth days. Mercury, too, spins slowly, but retains a weak field of its own. Both the Moon and Mars have none – at least today.

The core of a planet is not the only source of magnetic fields. Liquid water, especially salt water, can also generate energetic fields of a slightly different nature. To complicate matters, long-dead magnetospheres can leave behind faint magnetic patterns in ancient rock, which brings us back

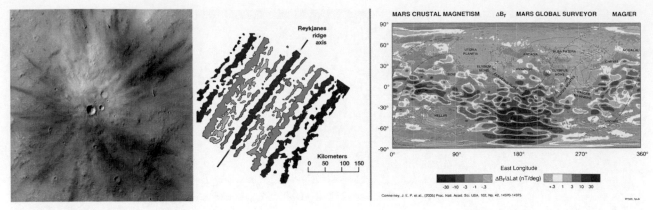

Fig. 1.7 Left: A new crater on Mars, spotted by the Mars Reconnaissance Orbiter (NASA/JPL/U of A). Center: Magnetic stripes on the sea floor south of Iceland show a mirror image on either side of the spreading ridge (after J. R. Hietzler et al.). Right: Some researchers suggest that the magnetic field patterns in Martian surface rocks may show signs of past plate movement (Image courtesy of NASA/JPL)

to Wegener and his plates. After his death, proof for Wegener's theory of moving plates came from an unexpected source, the science of paleomagnetism (the study of ancient magnetism in rocks). It also came from an unexpected location: the ocean floor.

In the 1960s and 1970s, scientists were able to chart changes in Earth's magnetic field. Their work – which incorporated rock samples from all over the world – suggested that Earth's magnetic field had reversed itself many times. Although scientists sometimes reverse themselves, it is rare to find a rock that does. Rather than a planetary magnetic field that was slowly losing energy, as some proposed, it now appeared that the field was simply going through cycles of reversal.

At the same time, aboard the H. M. S. *Owen* somewhere out in the Atlantic Ocean, researchers were dragging a device through the water, mapping magnetic stripes on the ocean floor. When they scrutinized the data, they realized that long, thin strips of rock on the sea bottom were strongly magnetized, while intervening strips were weak or had a magnetism that pointed the opposite way. The researchers were seeing a natural record of Earth's magnetic field over time. When rock becomes molten, any magnetic crystals inside are freed. The floating crystals align themselves with the magnetic field of Earth. When the rock solidifies again, the magnetized crystals become trapped in place, pointing in the direction of the magnetic field at the time the rock was molten. Later, as more molten rock comes up, it becomes magnetized in a new direction if the magnetic field has changed.

The critical clue came when scientists realized that the ocean rocks were a mirror image on either side of the Mid-Atlantic Ridge. It appeared that the ocean floor was spreading, leaving the same record on each side of the ridge, one side mirroring the other. All over the world, the magnetic record of continental spreading is written in the rocks.

Now, scientists have another planet with which to compare this phenomenon: Mars. The Mars Global Surveyor orbiter mapped similar magnetic stripes on the Martian surface. Faint repetitions of magnetic patterns seem to mirror each other. Though Mars does not have the same kind of moving plates that Earth has today, it may have had a similar process in the past.

THE GIANTS

The outer planets had strikingly different childhoods than their terrestrial siblings. While the young Sun was busily stripping the atmospheres from the inner planets, Jupiter, Saturn, Uranus and Neptune kept themselves at a distance. In the cold outer system, they were able to retain atmospheres similar to the earliest primordial cloud surrounding the Sun. Their atmospheres are dominated by hydrogen, a building block of all other gases, and helium. In addition to these major gases, Jupiter and Saturn have an abundance of ammonia, while the skies of Uranus and Neptune are seasoned by more methane. These poisonous brews rest above dense cores, but the giant planets have no solid surface. They are worlds of weather, spheres of tempest and thunderbolt, cyclone and rainstorm.

Jupiter and Saturn, the gas giants, are large enough that their interiors differ substantially from the other two giants. At their cores, pressures rise to the point where hydrogen becomes an electrified liquid metal. Densities in the lower regions of these behemoths are so great that laboratory studies of their conditions are practically impossible.[12] The cores of the other two giants, Uranus and Neptune, have lower pressures and are primarily water ice (hence their label of "Ice Giants"). The hearts of these outer worlds generate intense magnetospheres unlike those seen in the inner Solar System.

Although the giant planets are fascinating, human exploration will undoubtedly focus on the dozens of moons circling them. Many of these moons are as complex geologically as the terrestrials, with their own features echoing landscapes we have seen on the terrestrial worlds. Some are the size of planets, and one has more of an atmosphere than Earth.

This "new" Solar System, populated by planets the size of Mercury and smaller, begs the question: What about life out there on the moons? In the early 1960s, the planetary landscape looked fairly forbidding, as described by Patrick Moore:

> To sum up: in the Solar System, there are only two planets apart from the Earth which might possibly support life. On Mars there seems to be 'vegetation,' but intelligent beings are improbable. (Whether a Martian civilization used to exist and has now died out we do not know, but the evidence is rather against it.) Venus seems unsuitable for anything except, possibly, very lowly single-celled marine creatures. Yet can there be planetary systems circling other stars? The Picture History of Astronomy by Patrick Moore (Grosset & Dunlap 1961).

Back then, Moore and his colleagues didn't even consider the outer Solar System. But discoveries of extremophiles in deadly environments of Earth, along with the detection of subsurface oceans on several moons of the outer planets, have changed the game (see Chaps. 5 and 6).

Will human exploration be limited to the satellites of the outer worlds? Perhaps not. The outer planets themselves are currently beyond our technical capability to explore using manned ships, but eventually

12. Sandia Laboratories in New Mexico use their famous Z machine, a 130-ft-diameter torus X-ray generator, to simulate these horrific conditions, but the machine can only hold these pressures for fractions of a second at a time.

their skies may provide new provinces for human habitation. The idea of floating outposts or cloud cities is not new. In the 1726 satire *Gulliver's Travels*, Jonathan Swift described a floating city called Laputa. His "island city" floated by means of magnetism. Two centuries later, architect Buckminster Fuller proposed a mile-diameter geodesic globe that would retain buoyancy by the Sun's thermal heating.

Jupiter is a dangerous place to set up shop – floating or otherwise – with its high gravity and intense radiation. But Saturn, Uranus and Neptune offer tamer environments. The British Planetary Society's 1978 study of an interstellar probe – Project Daedalus – called for automated helium-3 fuel factories floating in the atmosphere of Jupiter, mining fuel for a massive spacecraft destined for the stars. Michael McCollum's science fiction novel *The Clouds of Saturn* described inhabited cities suspended in the Ringed Planet's clouds, held up by balloons filled with fusion-heated hydrogen.

A more technical NASA Glenn Research Center study presented in 2003 outlined the technical aspects of a floating Venusian outpost. Many of its concepts are applicable to the outer giants. Author Geoffrey Landis pointed out that although the Venusian surface bakes at a deadly 468 °C with an atmosphere 90 times as dense as Earth's at sea level, conditions at a 50-km altitude are far more Earthlike. There, a buoyant observation station could carry out remote-controlled exploration on the surface using robotic rovers, airships, etc. Eventually, even city-sized structures could sail the skies of the hothouse world next door. With modifications and changes in the gases used for flotation, even the gas giants could 1 day play host to such colonies (see Chap. 4).

How did the gas and ice giants grow to be the behemoths they are? How did they end up where they are now? The story is not as simple as it once seemed.

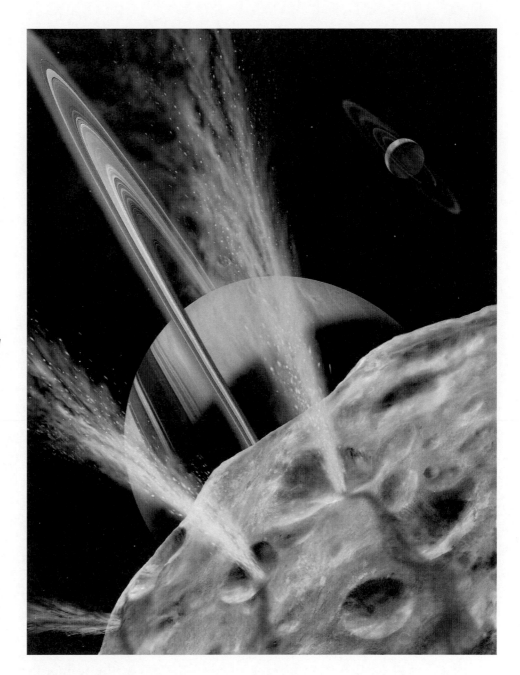

Fig. 2.1 Extended primordial rings encircle both Saturn and Neptune in this view of planetary migration in the early Solar System. According to the "Nice" theory, Neptune may have had several close encounters with Jupiter and/ or Saturn. In the process, wandering asteroids such as Saturn's outer moon Phoebe may have been captured, or even exchanged, from one planet to another. These captures would have been violent, perhaps causing internal pressures leading to volcanic eruptions, differentiation and surface collapse (Painting © Michael Carroll, original iteration courtesy Astronomy magazine)

Chapter 2
How *They* Got Here

Even science has its fairy tales.

Once upon a time, there lived a disk of dust and gas called the solar nebula. At the center of this pin-wheeling pancake cloud, laws of momentum and gravity dictated that material would fall in upon itself. As this central expanding mass grew heavier, its atoms could no longer hold themselves together, and nuclear fusion ignited the cloud into an incandescent orb. The Sun had arrived.

The fledgling starlet grew, pulling in more and more material with its increasing gravity. And just as an ice-skater pulls her arms in to spin more quickly, so the spinning of the disk began to increase. Irregular clumps of gas and dust grew into asteroids, comets, and ultimately planets. The rocky terrestrial planets established themselves near the Sun. Many of their volatiles burned away at the hands of furious solar winds gusting from the infant star. Out where it was colder and calmer, Jupiter and Saturn pulled hydrogen and helium to themselves from the solar nebula around them. Still further out, Uranus and Neptune were able to hold on to more water, becoming the ice giants we see today. In short, the planets formed where we see them now. This scenario is known as the Standard Model. It makes a nice story.

In science, however, models are meant to fall, and many researchers now believe that our planetary system's formative years were not so calm as we once thought. Dynamicists like Hal Levison of the Southwest Research Institute in Boulder, Colorado, have been studying the chinks in the armor of our beloved Solar System fairy tale.

The Standard Model faces two major problems. The first is what Levison calls the meter barrier. "Imagine these floating dust grains; these stick together and grow like dust bunnies. In objects 1–10 km size, if they hit gently enough they'll stick because of gravity. But you have this problem in between." Levison points to two coffee mugs on his desk. "If I bring these two cups together, no matter how gently, they aren't going to stick together. There is no good force to hold together things that are a meter to a decimeter size. Another problem is that there is aerodynamic drag on the particles. They are feeling the force of the gas they are moving through, feeling the headwind. That headwind will make them spiral into the Sun."

Very small particles can be suspended in the gas of the primordial solar nebula without settling. Large objects are unaffected by drag because of their mass. But those objects in between, in the meter range, should migrate into the Sun before they grow, effectively preventing any planets from forming. This puzzle has haunted dynamicists since it was introduced in the 1970s.

Some researchers propose that turbulence in the solar nebula may have concentrated meter-sized objects in eddies. The Sun's disk may have hosted swarms of these boulder-sized rocks, herded into vortices, gradually collapsing to form larger objects. Perhaps.

M. Carroll, *Living Among Giants: Exploring and Settling the Outer Solar System*, DOI 10.1007/978-3-319-10674-8_2, © Springer International Publishing Switzerland 2015

A second substantial change in our view of planetary evolution – and one that is better established – is that the giant planets did not form where they are today. "You cannot make Uranus and Neptune where they are," Levison says. "They must be closer in. The problem we find is that if you put a bunch of Earth-sized objects out in that region, they don't hit each other. They gravitationally scatter one another into Jupiter-crossing orbits, and Jupiter throws them out of the Solar System. You simply cannot get these guys to accrete [in the current Uranus-Neptune region]."

Out at the edge of our planetary system, the icy Kuiper Belt poses another mystery. In the outer Solar System, objects such as the ice dwarf Pluto move with high inclinations and high eccentricities in resonances – orbital relationships – with Uranus and Neptune. But because of the way planets develop, these objects had to have formed in circular, low inclination,[1] low velocity orbits. That's the only way to get planets to accrete, and yet observers see quite a different picture in the Kuiper Belt. Kuiper Belt Objects (KBOs) sail along on inclined, high eccentricity orbits. Multiple populations of KBOs seem made of the wrong stuff, as if they came from different locations superimposed on each other. The planets in their current configuration could not have formed where they are. Something had to change in the structure of our planetary system. Many dynamicists now suspect that the giant planets began in a compact configuration and then migrated out. This scenario is the heart of the Nice Model, named after the coastal town in France where Levison and three other scientists first combined forces to craft it.

The Nice Model proposes two extreme alternatives to the Standard Model. One scenario describes a smooth migration where Uranus and Neptune slowly spiral out through the solar nebula. The other idea has the four giant planets (and possibly a fifth that was later ejected from our Solar System) huddling close together until a global instability causes Uranus and Neptune to migrate wildly (see Fig. 2.2). Models show that their orbits cross each other, and even cross orbits of Jupiter and Saturn. Gravity from the two gas giants sends Uranus and Neptune packing, where they settle into a disk of planetesimals – a sort of distant asteroid belt – that no longer exists. This population of outer asteroids and comets, which would eventually give rise to the Kuiper Belt, was, in Levison's words, "a place where thousands of earth-like masses allowed things like Pluto to grow." It was the migration of the giant planets that destroyed it. But the disk's mass and its interaction with the worlds closer to the Sun essentially saved the planets, preventing them from being ejected from the Solar System.

The Nice Model's concepts are bolstered not only by observations of our own Solar System but also by studies of exoplanets, says Principal Scientist Dan Durda of the Southwest Research Institute in Boulder, Colorado. "A lot of what's advanced our thinking is seeing these hot Jupiters around other stars. It forced people into thinking, 'Oh, wait a minute, our simple picture of forming planets in a disk doesn't fit because how do you get a Jupiter in that close?' It forced people into thinking about what

1. Inclination refers to how tipped an orbit is to the "equator" of the Solar System, where most planets orbit around the Sun.

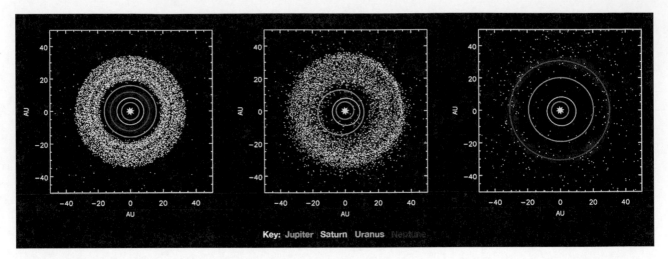

Fig. 2.2 The Nice Model. Left: The outer planets and planetesimal belt before Jupiter and Saturn reach a 2:1 resonance. Center: As Neptune (dark blue) and Uranus (blue-green) swap places, they scatter planetesimals into the inner Solar System. Right: Planets in their current orbits, after they have ejected the planetesimals from the Solar System (From R. Gomes, H. F. Levison, K. Tsiganis, A. Morbidelli, 2005)

Fig. 2.3 The discovery of "hot Jupiters," exoplanets orbiting near their stars, has played a key role in our understanding of how planetary formation occurs (Painting by the author)

happens in the disk as a big Jupiter is forming, and people like Bill Ward and others started realizing that there are all these interactions between the planet itself and the disk."

As a planet forms within a disk, researchers suspect that it sets up gaps and waves, and those density waves and gaps create torques on the planet itself.

Fig. 2.4 Saturn's rings provide insight into the dynamics of protoplanetary disks. In this Cassini spacecraft view, the tiny moon Daphnis orbits within the Keeler Gap, disturbing ring material in front and behind into waves and curls (Image courtesy of NASA/JPL/Space Science Institute)

When the planet forms and clears a gap, that gap can force the planet to move inward, Durda says. "It caused a whole decade of rethinking and re-understanding and learning about these processes that happen within the disks. It caused us to go back and look at a lot of the things that happened in our own Solar System, and that's what led to a lot of these gas/disk/planet interaction ideas. It's really been a paradigm shift from the idea that things have to form where we see them today to this idea of a whole lot of mechanisms that cause migrations and restructuring. It really is a huge shift in our understanding, and exoplanets have certainly played a key role."

Another contributor to our understanding of these processes has been the study of the structure of Saturn's rings. The Cassini mission has provided unprecedented detail of this complex, vast plain of material, offering scientists a real-life analog of a protoplanetary disk. Within the rings, gaps form at locations where particles are in gravitational resonances with large moons. Spiral density waves move throughout the system like the grooves on an old LP. Smaller moons, often embedded within the rings themselves, induce elegant ripples undulating for thousands of miles along the race-track-like bands of icy ring particles, while two-mile-high chevrons of material rear up at the edge of the B ring. "The Saturn ring system is probably one of the best natural laboratories for seeing the dynamics of a disk," Durda says.

Not everybody accepts the Nice Model. Its predictions seem at odds with some observations. Critics point out that the Nice Model is not confirmed by the number of craters and impact basins on Earth's Moon (which has less than what Nice calls for). The composition of fragments in the lunar samples points to asteroid impactors, while Nice suggests there should have been more comets falling on its surface. The Nice Model also suggests that when the planets migrated, their gravity would scatter asteroids into certain types of families that are not seen in the asteroids. This scattering would also have pounded the icy satellites of the outer planets to such an extent that they would have lost most of their icy crusts. This has not happened.

Computer studies demonstrate that a migration of Jupiter and Saturn would increase the eccentricities of the terrestrial planet orbits beyond their current values, leaving the Asteroid Belt with an excessive ratio of high to low inclination objects after the migration. In the case of the original Nice Model the slow approach of Jupiter and Saturn to their mutual 2:1 resonance, necessary to match the timing of the Late Heavy

Bombardment, can result in the ejection of Mars and the destabilization of the inner Solar System.

The separation of Jupiter's and Saturn's orbits caused by close encounters with one of the ice giants, called the "Jumping Jupiter Scenario," avoids these issues, but often results in the ejection of an ice giant. This has led some to propose an early Solar System with five giant planets, one of which was ejected during the period of instability.

Researchers continue to revise models that tweak the parameters of the Nice scenario and incorporate new findings. SwRI's Levison asserts, "No one has developed a model that is even close to being an alternative."

BEFORE THE NICE TIMES – A DIGRESSION

Scientists of the eighteenth century had two clues about the formation of planets. The first was that stars spin (as could be seen by sunspots on our own Sun and the regular changes in light from distant stars). The second clue was that all the planets travel in nearly circular orbits – and in the same direction – around their parent star. This led to the early conclusion that the Sun and planets issued from a disk of material that was spinning. We now call this disk the solar nebula.

Immanuel Kant, a Prussian philosopher and writer, first put forth the concept of a solar nebula in 1755[2] (and independently, in 1796, it was put forward by French astronomer Pierre Simon Laplace). Kant proposed that clouds of gas floating through the universe would be unstable and would tend to clump together gravitationally. As these cosmic clouds collapsed, Kant suggested, they rotated. This rotation would spin the clouds into flattened pancakes, leading ultimately to planets circling a central star.

Others have put forth other ideas. In 1905, geologist Thomas Chamberlin and astronomer Forest Moulton proposed that a star had passed close to the young Sun, pulling material away into a spiral arm. This arm coalesced into the planets. The advantage to their theory was that the thickest part of the arm would have been in the center, where the gas and ice giants are, with the thin parts leaving small terrestrial planets on one end and comets and distant asteroids on the far end. At the time, spiral galaxies were thought to be stars with arms of material pin-wheeling out from the center, and these were seen as evidence for their theory. Three decades later, Soviet astronomer Otto Schmidt suggested that the Sun drifted through a dense cloud of interstellar gas, trailing debris from which the Solar System eventually formed.

Other theories came along, too. It turns out that Immanuel Kant was right, but no one actually saw disks around stars until the advent of radio telescopes, which could peer into visually opaque dark clouds. More recently, the Hubble Space Telescope (HST) and Spitzer telescope have discovered dozens of planet-forming disks around stars, confirming Kant's original scenario.

2. Kant may have been influenced by the earlier work of Emanuel Swedenborg.

Fig. 2.5 Protoplanetary disks in the Orion Nebula, called proplyds, provide confirmation of Immanuel Kant's theory of Solar System formation (HST image courtesy of STScI)

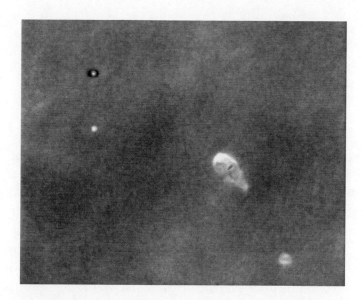

BEYOND THE SNOW LINE

The regions nearest the Sun were far too hot for water or gas to condense out as ice. Planets there pooled into silicate, rocky bodies. The hot inner Solar System was adrift in a fog of tiny igneous spheres called chondrules. Chondrules have been found in many ancient meteorites, primordial flotsam born in very high temperatures (1,500–1,900 K). These chondrules stuck together, becoming the terrestrial planets. Farther out from the great

Fig. 2.6 Spherical chondrules are composed primarily of silicate minerals such as olivine and pyroxene. Found in many kinds of meteorites, they lend clues to the early Solar System's development (Photo courtesy of Dan Durda. Used with permission)

disk of the solar nebula, temperatures were cold enough to form ice. The line between the two regions is called the "snow line." Beyond that line, smaller bodies assembled themselves into balls of ice and rock. The gas giants, with their immense gravity, pulled hydrogen and helium directly from the nebula around them. They may have grown steadily as their gravity increased, or they may have benefitted from waves of material cast toward them through gravitational instabilities in the cloud. Either way, the gas giants ended up as massive globes of hydrogen, helium and ammonia with metallic cores.

Whether the Nice Model is correct or not, we know that Uranus and Neptune formed beyond the snow line, too. Although Jupiter and Saturn were able to collect massive amounts of hydrogen and helium, the ice giants have ended up with less of those gases and more gases associated with frozen water and organics: oxygen, nitrogen, methane and carbon.

While they were at it, all those spinning gas and ice giants cocooned themselves within their own disk-shaped clouds. Like the infant Sun's inner system, the central cloud that led to the gas giants was warm, heated by the energy of accretion.

As Jupiter and Saturn cooled, they shrank, leaving behind a cooling cloud of gas, ice and dust. Within these mini accretion disks, ring systems formed and, farther out, moons coalesced. Jupiter's four major moons, the Galilean satellites, provide the perfect example of the result. Close in to Jupiter, Io and Europa collapsed into spheres with dense, large rocky cores and, in the case of Europa, a relatively thin water-ice crust. These are Jupiter's version of the terrestrial planets. In Jupiter's outer cloud, where there was more water, Ganymede and Callisto formed as larger, less dense worlds with small stony centers and deep ice crusts. Icy giants. (Oddly, moons of the ice giant Neptune do not seem to follow this organized pattern of decreasing density with increasing distance, perhaps because of the large interloper Triton, which came from outside of the system.)

As the formation of a gas or ice giant comes to an end, the planet will continue to accumulate gas, rock and ice from the solar nebula – the cloud of dust and gas around the Sun. The disk of material, orbiting the giant planets in their equatorial planes, begins to ebb and flow with the same kinds of eddies that initially cause the planets in the Sun's own cloud. According to new studies,[3] as the protosatellite cloud coalesces around its parent planet and moons form within, the gravity of the newborn moons disrupts the cloud, triggering spiral waves. As the satellites grow, the effect becomes more marked, so that the moons' orbits begin to spiral in toward the planet. As more material flows into the cloud, the inner satellites move in toward the planet while new ones are born toward the outside of the gaseous disk.

Dan Durda describes the early evolution of the vast satellite system at Jupiter: "The implications are that the four Galilean satellites we see are the last of a whole conveyor belt of satellites that formed and got gobbled up

3. For more, see www.swri. org/9what/releases/2006/ canup.htm.

by the planet as they migrated in through this disk." As the young Sun matures, Durda, explains, it endures the T-tauri phase, an energetic stage in which its solar wind blows most of the dust and gas from the planetary system. "What shuts off that disk and shuts off that conveyor belt is the T-tauri phase when the Sun really does turn up and clears the disks – not only the protoplanetary disks but the protosatellite disk around Jupiter as well. It gets rid of those gas disks and shuts off that process."

This conveyor-belt mix of satellite birth and destruction keeps the mass of the moons at a constant total. The satellite systems of Jupiter, Saturn and Uranus are quite different from each other. Jupiter's four Galileans are nearly alike in size, whereas Saturn has one giant moon along with many mid-sized ones. The satellites of Uranus are somewhat comparable in arrangement to those of Jupiter's. Even so, researchers point out that Jupiter, Saturn and Uranus do, in fact, have a similar ratio between the mass of the planet and the overall mass of the satellite system, with the satellites making up roughly one hundredth of 1 % (0.0001) of the mass of their parent planet.

The outer Solar System eventually settled into the arrangement we see today. It is a realm of bitter cold, mind-numbing isolation, and lonely darkness. But the Solar System's outermost planetary regions play host to some of the most spectacular formations, beautiful landscapes and bizarre phenomena ever witnessed by humankind. It is waiting for us, but before the humans set foot on those cold and remarkable shores, the robots must go forth. Some already have.

Fig. 3.1 The Galileo
spacecraft, with its jammed
main antenna, on final
approach to Jupiter in 1997.
As it approached the Jovian
system, the craft made a close
pass by the volcanic moon Io
(Painting © Michael Carroll)

Chapter 3

How *We* Got There

Our Solar System seems to be arranged in a strange way. The inner planets huddle near the Sun, circling fast in their tight orbits like vultures over Serengeti leftovers. But beyond Mars, everything spreads out. If our Sun were the size of the head of a pin, the four inner planets would circle within a region 22 in. from it. But on this scale, Jupiter's orbit would keep the planet at 66 in., and Neptune would revolve around the pinhead Sun some 39 ft away. The distances to the outer planets seemed, to early spacecraft designers, a great chasm. Mars was just barely doable; the Red Planet could be reached in 6 months on a fast track. But a probe to Jupiter faced years in the harsh vacuum of space, and a voyage to Neptune promised the daunting agony of a long decade.

Distance and time were not the only concerns. A vast donut-shaped cloud of asteroids circles the Sun just beyond the orbit of Mars. By 1961, observers had charted 2,000 ice balls and rocks ranging from Ceres (950 km in diameter) down to what was then the limit of resolution. Thousands more sightings would soon follow. How many more were there, lurking just below the visibility of our telescopes? Would these flying mountains of rock and metal afford a permanent, deadly barrier to any craft trying to reach the outer Solar System?

Flight engineers knew from experience that it would take a mere speck of sand to cripple a spacecraft. On August 4, 1969, the *Mariner 7* spacecraft had an appointment with the planet Mars. It was an appointment that the little robot would nearly miss. Its sister craft, *Mariner 6*, was targeted for an equatorial flyby a few days earlier, on July 30. *Mariner 7* was to follow with a reconnaissance of the south polar region. Only the *Mariner 4* had successfully returned images of Mars 4 years before, and scientists were eager to see more. But just 5 days out from Mars, and 7 h before *Mariner 6* made its closest approach, NASA's deep space tracking station in Johannesburg lost contact with *Mariner 7*. The flight team leaped into an emergency search with any antenna large enough to pick up the wayward craft's signals.

Just minutes before *Mariner 6* passed the Red Planet, California's Goldstone Tracking Station picked up a faint signal from *Mariner 7*. Controllers commanded the prodigal craft to switch to an antenna with a wider beam, and flight engineers were able to regain control of the spacecraft in time for its polar encounter a few days later. But the report they heard from the spacecraft's onboard systems was troubling. The robot's Canopus star tracker had lost its lock with the star, allowing the spacecraft to turn away from Earth. Something had apparently bumped the spacecraft, and it wasn't a gentle bump. From 20 to 90 telemetry channels had been lost, and *Mariner 7*'s velocity had changed. Gas appeared to be venting from one of its tanks. It was possible that one of the probe's tanks had ruptured, but the combined evidence pointed to an impact by a micrometeroid. The event gave spacecraft designers pause, to say the least. After all, if a spacecraft could be damaged by micrometeoroids at Mars, what would happen in the Asteroid Belt beyond? The event would haunt designers for years to come.

M. Carroll, *Living Among Giants: Exploring and Settling the Outer Solar System*, DOI 10.1007/978-3-319-10674-8_3, © Springer International Publishing Switzerland 2015

THE PIONEERS

In 1972 and 1973, *Pioneers 10* and *11* left the launch pad bound for Jupiter. They were unlike anything that had taken to the skies. Rather than gangly solar panels or coat-hanger antennae, the craft were dominated by a large dish antenna for long-distance communication. By today's standards, the probes were small and simple. Weighing in at just 258 kg, their main "bus," or body, was less than half the mass of the next Jupiter missions (the Voyagers). Their main antennae spanned a diameter of less than 2.7 m, as compared to the advanced Voyagers, whose main antennae would be 3.7 m across.

Pioneers were spin-stabilized; rather than keeping attitude control with small thrusters, the craft spun like a top to remain stable. This meant that instruments were in constant motion, even the imaging systems. The Pioneers carried no cameras, but instead had "imaging photopolarimeters" from which fairly low resolution images could be reconstructed. The major reason was simple: a spinning spacecraft needed to carry less fuel, and no mission of this length had ever been attempted before. The Pioneers were the first to use nuclear power.

Rather than a detailed reconnaissance, the primary goal of the mission was one of survival, remembers former director of JPL Dr. Ed Stone. "*Pioneer 10* and *11* had two key long-term objectives. One was, *can we get through the Asteroid Belt? And two was, how serious is the radiation around Jupiter?* Both of those were key in our mission design." Even so, the robots carried 11 of the most advanced experiments of the time to study the magnetic fields of interplanetary space and the environs of Jupiter and Saturn. They mapped the solar winds and how they interacted with the planets, measured temperatures of the atmospheres of the gas giants and some of the larger satellites, and snapped spin-scan images of the planets and largest moons.

Perhaps most important for future explorers – both robotic and human – the spacecraft nailed down the radiation levels surrounding each planet. The energetic fields and particles surrounding planets create a sort of magnetic bubble, or shield. These bubbles of magnetic fields, called magnetospheres, protect a planet from the Sun's solar wind, but they also build up a shell of deadly radiation. Although a spinning spacecraft was not ideal for imaging, it was perfect for scanning the complex structure of a planet's magnetosphere.

Ed Stone's expertise lay in planetary magnetospheres. "I remember being surprised when *Pioneer 10* was approaching Jupiter, because we had scaled the size of the magnetic field from that of Earth. It's a pressure balance between the field inside and the solar wind outside. We knew what the solar wind was; we'd measured it. We thought we knew what the magnetic field was from ground-based radio observations, and that said that the magnetosphere should be about 40 Jovian radii out (in the case of Earth, it's about 10 Earth radii out). When *Pioneer 10* first detected the magnetosphere it was over 100 Jovian radii."

Both Pioneers proved incredibly successful. They returned the first close-up images of Jupiter and the first good images of its polar regions. Both craft also imaged the Galilean satellites of Jupiter, and *Pioneer 11* ventured on to investigate Saturn and its moon Titan. Because of the spin-

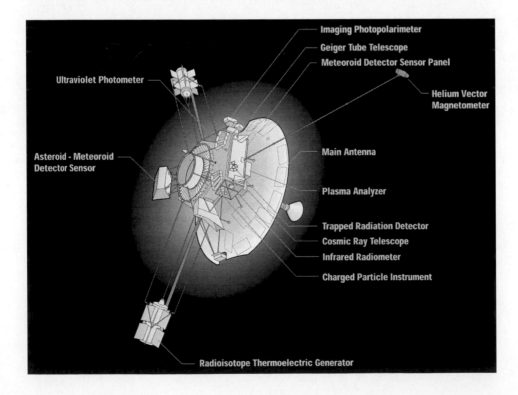

Fig. 3.3 Pioneer 10 and 11 carried an array of scientific instruments powered by radioisotope thermoelectric generators (Image courtesy of NASA)

ning nature of the spacecraft, the images were built up of thin strips scanned as the spacecraft turned. Because a typical color photograph[1] took 30 min to scan, much detail was lost in the long assemblies of imaging. Still, the twin craft gave us our first hints at the nature of the moons of the giant worlds (see Chap. 5, Fig. 5.5).

The Pioneers returned the first in situ measurements of micrometeoroids in the interplanetary space beyond Mars, significant because no spacecraft had ventured through the Asteroid Belt, and some estimated

1. For example, *Pioneer 11* image # G3 of a crescent Saturn took 28 min for two images using different filters.

Fig. 3.4 *The twin Pioneers' scanning cameras imaged polar views of Jupiter (top) and Saturn in detail never seen before (NASA/JPL images processed by Ted Stryk)*

that micrometeoroids would cripple any passing spacecraft beyond survival. The twin craft made the first direct measurements of radiation fields around gas giant worlds (see Chaps. 4 and 5). Among many "firsts," *Pioneer 11* also discovered Saturn's famous F-ring.[2]

THE VOYAGER PROJECT

The instrumentation of the small Pioneers whetted the appetites of the planetary science community for more. Emboldened by the successful crossings of the Asteroid Belt, NASA resolved to dispatch two larger, more advanced craft to the outer Solar System. The timing could not have been better. At the end of the 1970s, the outer planets aligned in a rare arrangement. Jupiter's location enabled it to be used as a gravity slingshot to the other planets of the outer Solar System, cutting flight times by many years. This alignment would not occur again for another 175 years.

NASA had been lobbying for a sophisticated pair of Grand Tour spacecraft called the Thermoelectric Outer Planets Spacecraft, or TOPS. The nuclear-powered spacecraft would have weighed in at over a ton, bristling with multiple atmospheric probes to drop off at all four of the giant worlds. The first craft would launch to Jupiter, bank off to Saturn, and sail by Pluto for good measure. The second craft would round out the giant planet reconnaissance by using Jupiter's gravity as a boost to Uranus and Neptune. TOPS would be equipped with the most advanced science money could buy. But the latter was the problem. Congress balked at the price tag. Instead, NASA looked to its highly successful Mariner spacecraft, retooling it with a huge antenna and twin radioisotope thermoelectric generators. Christened Voyager, the smaller craft cost a third of what TOPS would have.

Engineers fashioned two Voyager spacecraft. *Voyager 1* launched on a trajectory that would carry it past Jupiter and on to Saturn. Rather than continuing on to Pluto, which some doubted could be reached before the spacecraft failed, an emphasis was given to a close study of Saturn's foggy moon Titan. If the first spacecraft survived, *Voyager 2* would be targeted to do a scaled-down version of the Grand Tour, passing Jupiter and Saturn before traveling on to Uranus and Neptune over the course of a long decade.

The twin Voyagers were already under construction at the time of the Pioneer Jupiter encounters. That, says Ed Stone, was fortunate. "We benefitted a lot from the Pioneer encounters, because we learned that the radiation environment was much worse than our models suggested, which meant we took nine months off at the beginning of 1974, redesigned circuits, and replaced parts, because we were concerned we would not survive the Jupiter encounter."

2. *Pioneers 10* and *11* would also be the first objects destined to actually leave our Solar System.

Fig. 3.5 Left: The Voyagers were powered by three plutonium-energy RTGs (radioisotope thermoelectric generators), seen just behind the white metallic shield. Note the huge white high-gain antenna affixed to the top of the spacecraft. Right: Voyager's scan platform, on the boom at left of frame, carried a host of instruments, including the telescopic cameras (Photos by the author)

Armored with improved shielding and stuffed with advanced electronics, survive they did, and in grand style. The first mission discovery concerned those nasty magnetic fields that flight engineers were so apprehensive about. The true nature of those radiation fields, Ed Stone observes, was "like nothing we had been thinking. Nobody had any idea what was ahead. And then it went away! And then it came back, because the magnetic equator of Jupiter is inclined, so as Jupiter rotates, it goes up and then down, in and then out, like a big, floppy disk. It's a very flat disk that is wobbling as Jupiter rotates, and so all of a sudden it wasn't there, and then there it was again. And that was just a precursor of the kinds of discoveries Voyager was going to make, just time after time. Suddenly things just changed your whole perspective. Your terracentric view was just that; it was based on things we thought we knew about Earth, and we extrapolated all that, but it turned out that there was a much broader set of objects out there than our terracentric experience prepared us for."

Despite the Voyagers' improved shielding, the craft barely weathered the radiation, says Ralph McNutt, chief scientist in the Space Department at the Johns Hopkins University Applied Physics Laboratory. "What people don't realize is that *Voyager 1* took a real beating on its traverse through the [Jupiter] system. There were something like twenty power-on resets of the main computer due to energetic electrons hitting the main control and command computer during closest approach. We knew that there were large fluxes of electrons, and as a result they added an extra thickness of copper around the electronics box to try to beat back some of the radiation. That's probably the reason that some of the Voyager observations are as good as they are, because of that extra shielding. It really was with the Voyagers that we finally got a handle on how nasty the radiation environment was."

Unlike the Pioneers, the Voyagers were stabilized on three axes, using specific stars to guide them in three-dimensional space. This enabled them to point cameras and other instruments precisely, using a scan platform.

The spacecraft were even able to take long exposure images, something a spinning spacecraft could never attempt before this. Both Voyagers returned spectacular views of the Jupiter and Saturn systems. Ed Stone remembers trying to keep up with the flood of new information. "Every day there were discoveries, discoveries we hadn't really imagined that we would be making. That's what made it such a mission of exploration, of discovery. We knew we were going to learn things, but we had no idea how diverse the system was, or how many different and interesting objects we'd be seeing. Every one of those moons was different. They could have all been the same, all heavily cratered ancient surfaces. Points of light, that's what they *were*. And suddenly they became real worlds, each one separate and unique."

Voyager discoveries included the rings of Jupiter, previously unknown rings of Saturn, many small moons, volcanoes on Io, probable under the surface seas found on Europa, possible evidence for geologic activity on many ice moons, including Europa, Enceladus and Dione, and a complex, dense atmosphere on Saturn's giant moon Titan. Many of the bizarre landforms seen on the moons came at the hands of tidal heating, a force of nature that was underestimated before Voyager. Icy moons expert William McKinnon comments, "There had been predictions of tidal heating and what it could do, but we'd never seen anything like it, and no one had a real appreciation of its power. Even tiny worlds are very active."

"Before Voyager, for example, the only known active volcanoes were here on Earth," Ed Stone adds. "And then we flew by Io, with ten times more volcanic activity than Earth. The only known liquid water ocean? Right here on Earth. We flew by Europa. The only known nitrogen atmosphere? Here on Earth. Then we flew by Titan. Time after time, things which we thought we knew turned out to be not so. The magnetic fields of the planets should be lined up with the rotation of the planet. Guess what happened at Uranus? It was aligned more with the equator. Same thing is true of Neptune. So time after time, our terracentric view was not nearly broad enough for our own Solar System."

That trend continued even further into the mission. *Voyager 1* sacrificed any encounters beyond Saturn in order to fly close to Titan, but *Voyager 2* continued its Grand Tour, coasting through the Uranus system in 1986. The craft imaged Uranus' gunpowder-black moons and rings, mapped subtle storms and cloud bands and charted its bizarre offset magnetic field.

The weather on Uranus turned out to be far less energetic than expected. The belts and zones so familiar on Jupiter and Saturn faded into near-invisibility in the green Uranian smog. Only a handful of distinct storms could be seen in images, says outer planets' expert Heidi Hammel:

> The Uranus flyby was just a flash in the pan, a brief moment in time, that happened to take place in a very special time for Uranus. It was there at the southern summer solstice. Uranus is tilted; it lays on its side. That means its seasonal variability is

Fig. 3.6 A pre-Voyager view of the spacecraft flying by Uranus and its small moon Miranda. At the time, investigators expected to see distinct banding on the giant world (Painting by the author)

Fig. 3.6 A pre-Voyager view of the spacecraft flying by Uranus and its small moon Miranda. At the time, investigators expected to see distinct banding on the giant world (Painting by the author)

extreme: at the solstice one pole is illuminated and the other is in utter darkness. So when Voyager flew by, it happened to fly by when the southern pole was pointed almost directly at the sun. No matter how much the planet spins, you're still looking at the pole. At that time, Voyager saw pretty much no discreet clouds; it saw ten little features. [Uranus] had a thick, cloudy atmosphere but it was uniform. So you had this solid cyan-colored ball with just occasional faint markings that could be tracked. And it was all in the southern hemisphere because that's all that was illuminated. So that's pretty boring compared to Jupiter with its beautiful swirling belts and zones.

The green giant's boring countenance would change, years later, with ground-based imaging (see Chap. 4).

Voyager 2 opened up the realm of the ice moons to the scientific world. The moons of Uranus bear similar diameters to those of the mid-sized satellites of the Saturnian system. The Voyagers had hinted at past geologic

Fig. 3.7 *JPL's control center for the Voyager spacecraft, at the time of the Saturn encounters, ca. 1981 (Photo by the author)*

activity on several of Saturn's moons, including Enceladus and Dione, and *Voyager 2* spotted the fingerprint of past excitement at Uranus as well. Particularly, Miranda showed a face uplifted and twisted by forces within or outside of it, and Ariel and Titania both hinted at past epochs of cryovolcanic activity (see Chap. 8). Light levels at Uranus are quite low – about 1/400th that at Earth – and the spacecraft was traveling at over ten times the speed of a rifle bullet. Adding to the challenge was a balky scan platform. The aiming system on Voyager had jammed in one direction during the Saturn flyby, so instead of moving the camera, the entire vehicle had to be slewed to compensate for long exposures to prevent image smearing. Flight engineers did a fantastic job under the most difficult of circumstances. It was good practice for the next encounter, 3 years later. Neptune awaited.

As Voyager approached the last of the ice giants, planetary meteorologists began to scratch their heads in earnest. What was all the activity? After all, Neptune was farther away from the Sun than any of the other giant worlds, and would be receiving the least energy. Would it not be the most quiet?

As the images became more clear, one planetary weather theory after another fell by the wayside. Neptune had the fiercest winds of any world, with cobalt jet streams crowned by glistening white methane cirrus clouds. A great blue storm, reminiscent of Jupiter's Great Red Spot, rotated into view. The spinning cyclone could swallow an entire Earth. And like the satellites of Uranus, Neptune's star was a moon, the large Triton.

Ed Stone sums up the decade-long planetary missions in this way: "I like to think of bookends for the planetary phase. Io was the first, with its volcanoes. There were a whole series of discoveries that came along, with Titan and others. Then the other bookend for the planetary phase was Triton, which is 40 Kelvins, 40 degrees above absolute zero, and there are

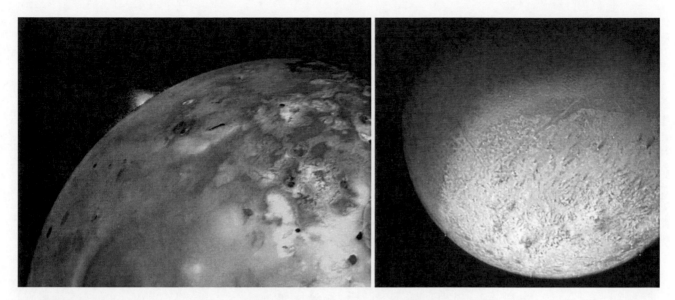

Fig. 3.8 Ed Stone's "bookends"
of the Voyager missions: Io
(left) and Triton (Images
courtesy of NASA/JPL)

geysers erupting from its polar cap. So that's a nice bookend. These things
with the kind of activity that you just wouldn't have imagined, and all the
things in between."

The Voyagers continue to broadcast from locations at the very edge of
the Solar System. As of this writing, Voyager 1 appears to have passed
through the edge of the sun's influence – the heliosphere – and into true
interplanetary space. But the logical step after flyby missions are robots
that can do long-term reconnaissance, and at Jupiter, this was the task of
the Galileo orbiter and atmospheric probe.

THE GALILEO ORBITER/PROBE

After the sophisticated Voyager flybys of the gas and ice giants, the survey
of the outer Solar System was complete. But it was a survey only. To begin
to understand detailed trends of the giant worlds and their attendants,
researchers needed a long-term look, and that look would have to come
from an orbiter.

In 1995, the massive Galileo orbiter arrived at Jupiter. It combined the
best of Pioneer and Voyager, with a spinning section to study fields and
particles, and a de-spun section to scrutinize the clouds and rings of Jupiter
and the surfaces of the Galilean satellites and other moons. But before the
craft arrived, crisis struck.

The main antenna, designed to open like an umbrella, refused to
deploy. The problem was the result of a longer-than-planned cruise
through the hot inner Solar System. Galileo had been designed for a fairly
rapid cruise out to Jupiter, but with the loss of the *Challenger* space shuttle
in 1986, Galileo's powerful liquid propellant upper stage was deemed too
dangerous a cargo for the shuttle bay. Instead, flight designers substituted

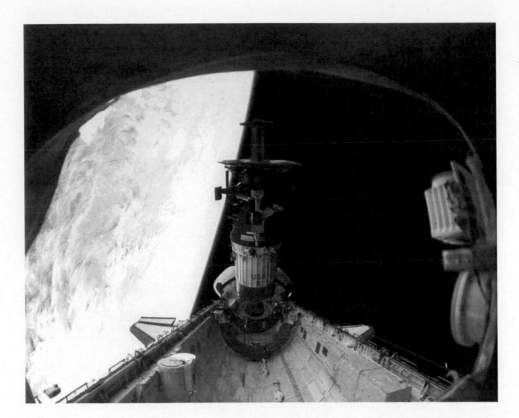

Fig. 3.9 The Galileo orbiter and probe begin their long journey to the outer Solar System, deployed by the space shuttle Atlantis on October 18, 1989 (Image courtesy of NASA/JSC)

a safer but less powerful solid propellant stage. This necessitated a leisurely 6-year looping voyage past Venus and twice past Earth before the craft could gain enough speed to set out for the Jovian system.

En route, the craft carried out the first flyby of an asteroid, 951 Gaspra, and another asteroid flyby of 243 Ida, discovering its small moon Dactyl. During this time, lubricants at the ends of the antenna's ribs migrated, and it jammed. By the time the huge orbiter arrived at Jupiter, its 4.8-m diameter antenna was crippled and useless, and the stream of data destined for the distant Earth had to be sent in a tiny antenna the size of a coffee can. But programmers and engineers demonstrated their typical creativity, rose to the challenge, and taught the craft to compress data not only in the way it transmitted it but also in the way it gathered and interpreted it. On the ground, adjustments were also made at receiving stations, so that the mission could still achieve an impressive 70 % of the mission's multiyear goals.

Five months before the spacecraft rocketed into orbit around the king of worlds, it deployed the Galileo Atmospheric Probe, which was destined to enter Jupiter's atmosphere. The early release gave the orbiter enough separation from the probe to safely miss Jupiter and go into orbit. The probe systems were controlled by a simple timer, and powered up some 6 h before entry. Entering Jupiter's atmosphere on December 7, 1995, the Galileo probe survived entry speeds of over 106,000 mph, temperatures twice those on the surface of the Sun and deceleration forces up to 230 times the force of Earth's gravity. At the speed the probe was traveling

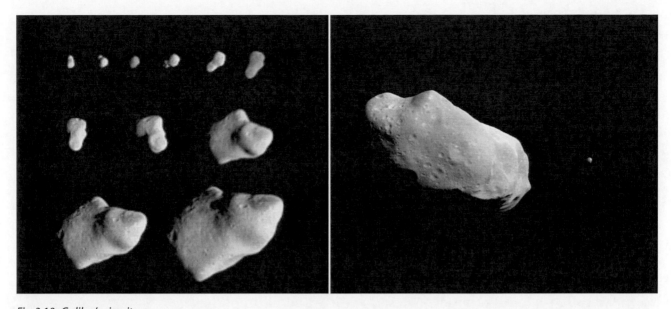

Fig. 3.10 Galileo's circuitous cruise included the first two close flybys of asteroids. The craft encountered Gaspra in October of 1991 (rotation sequence at left) and Ida in August of 1993. Note Ida's tiny moon Dactyl, just 1.4 km across (Images courtesy of NASA/JPL)

during entry, it would have crossed the distance from Los Angeles to San Francisco in roughly 10 s.

After the heat of entry died down, a small drogue parachute deployed, pulling the back shell from the probe and dragging out the main parachute. Ten seconds later, Galileo dropped its protective heat shield. Just 1½ seconds later, with the heat shield some 27 m away, the official mission commenced, and Galileo began to investigate the banded skies of Jupiter. It relayed data obtained during its 57-min descent back to the Galileo orbiter more than 209,215 km overhead. Real-time data had to be stored for later playback because of the orbiter's failed main antenna.

Fig. 3.11 Left: Nearly half of the mass of the Galileo probe's aeroshell (heat shield) was blasted away during its high speed entry into Jupiter's atmosphere. Center: The probe itself, with all of its scientific instruments, was pulled from the aeroshell by a parachute. Right: Wind tunnel test of the Galileo probe parachute (Images courtesy of JPL/NASA)

Engineering data from the probe indicates that it was not smooth sailing. Galileo began transmitting 53 s late due to a wiring problem. The probe swung back and forth beneath its parachute every 5 s, and was apparently spinning once each 20–25 s. The 400-mph winds buffeted the lonely emissary as it fell through Jovian canyons of clouds. Its parachute may have been damaged, as descent speeds were higher than anticipated, and the accelerometer designed to trigger deployment of the chute appears to have been installed backwards, making parachute deployment a matter of pure good fortune.[3] The probe watched for lightning, which was surprisingly rare. In addition to water and other constituents, Galileo monitored ammonia, which is associated with dramatic updrafts in active areas.

However, Galileo fell victim to the cosmic version of Murphy's Law (i.e., whatever can go wrong will go wrong). Although mission planners hoped to sample an average place on Jupiter, the probe happened to descend through a most unusual hot spot, where the all-important clouds were rare. The probe was packed with cloud-measuring instruments, but those instruments nearly starved. The probe had passed through very few high-altitude ammonia clouds, and even fewer water clouds. It was beginning to look as if all the models of Solar System evolution were wrong.

APL's Ralph McNutt cautions that atmospheric probes have limitations. "We dropped off the Galileo probe, and we had some period of data. We didn't quite know what happened to the purported water, and wondered if we went into the wrong place, which is the problem with a drop-in probe. It's the easiest to do, but it also means you're looking at a very localized area, so you don't have a good feel for what's going on globally."

As the probe continued to gather its precious data, internal temperatures increased to dangerous levels. During its fiery entry, technicians believe hot gas entered into the shell around the probe, perhaps through a breached seal. Experiments were calibrated for certain temperature ranges, and the probe's temperature quickly rose above those levels. This created challenges for scientists who tried to interpret the numbers coming back from Jupiter. For a few days, scientists struggled to understand what had happened. Through herculean efforts, analysts untangled the data, and the atmospheric probe's legacy remains a turning point in our understanding of planetary atmospheres and weather.

The Galileo orbiter carried out 35 circuits around Jupiter over the course of nearly 8 years. Each orbit was a miniature rerun of the Voyager flybys, involving spectacular encounters with Jupiter and multiple moons, combined with continuing studies of the structure of the magnetosphere and surrounding interplanetary space environment. At the end of its long mission, controllers commanded the craft to impact into Jupiter, where its destruction assured that the spacecraft could not eventually contaminate Europa or any other moon that might be a host of in situ life.

3. A similar problem with the installation of a parachute accelerometer caused the near-destruction of the return capsule on the Genesis solar wind mission, which crash-landed in the Utah desert in 2004. Genesis was, nevertheless, a successful mission.

CASSINI AND HUYGENS

In 2004, it was Saturn's turn for an orbiter. Like Galileo, the international Cassini spacecraft consisted of an orbiter and atmospheric probe, but its probe was designed to explore Saturn's planet-sized moon Titan.

The 6-metric-ton Cassini orbiter settled into orbit around Saturn after a 96-min burn of its main engine. Just 150 days earlier, Cassini had gently ejected the European Space Agency's (ESA's) Huygens probe on a slightly different path that would intersect Titan. The separation enabled Cassini to relay data from the probe to Earth in real time as the Huygens descended through the dense atmosphere of Titan, much as Galileo had done with its Jupiter atmospheric probe.

The school-bus-sized Cassini orbiter is the most complex spacecraft ever flown. Its formidable suite of advanced instruments listens to the music of Saturn's magnetic fields, images the varied moon surfaces and rings, and charts the aurorae at Saturn's poles.

Although the craft scrutinizes many aspects of the Saturn system, its primary goals are twofold: to study Saturn and to study Saturn's planet-sized moon Titan. The majority of Cassini's tracks around Saturn bring it close to Titan. As it flies by, some instruments can image the surface through the haze, while its entire battery of experiments studies atmospheric haze layers, clouds and surface features. Cassini also beams radar through the atmosphere to map the surface in great detail, much as the Magellan did at Venus in the 1990s. Finally, when Cassini passes behind the moon as viewed from Earth, researchers monitor the changes in its signal to chart the structure of the atmosphere. This is also done with Saturn itself, and the same technique is used to search for rarified atmospheres around other moons and to map the density of the ring system.

Fig. 3.12 The Cassini orbiter deploys the ESA's Huygens Titan probe in this artist's rendition done for JPL before the actual launch (Image courtesy of NASA/JPL; art by the author)

Cassini's precious partner, Huygens, arrived at the upper fringes of Titan's air on January 14, 2005. The probe nested within a 9-ft diameter heat shield. Because of Titan's low gravity (one-tenth that of Earth's) and dense atmosphere, the probe required much less protection than that of Galileo. Titan's low gravity and extended atmosphere make it unusually easy to land on. Engineers estimated that Huygens would enter the atmosphere at a speed of 13,000 miles per hour, experiencing 12 G's of deceleration (compared to Galileo's 350 G's).

Huygens was a spectacular success. The probe initially descended on a large parachute to slow its descent, but later ditched this for a smaller parachute, enabling it to get to the lower atmosphere more rapidly to coordinate its descent with the flyover of the orbiter. Huygens was designed to spin, so images from its DISR imaging system – essentially a slot camera – could be assembled into 360° panoramas.

Huygens returned spectacular images from high altitude as well as from the surface after surviving its landing. "For those initial images of the surface, I was back in Germany with the inner circle of people who first saw these images," says project scientist Peter Smith, a veteran of many planetary missions. "They were just astounding. It looked like the coast of Italy: streams coming down a hill into what looked like a dry lake bed and things that looked like a fault line that had shifted. Of course we know that the rain is methane and the rocks are water-ice and there are hydrocarbons all over the place, but still, the shapes of the surface looked awfully familiar. That was a big surprise to me."

Shortly after the spectacular triumph of Huygens, Cassini began logging a series of groundbreaking discoveries, including the bizarre tectonics on Dione and Tethys, the baffling equatorial ridge of Iapetus, the cryovolcanic geysers of Enceladus, numerous waves, spikes, and other phenomena in the rings and the unveiling of a host of previously unknown moons. Its studies of Titan complemented those of Huygens, mapping vast dune seas and charting the methane rainstorms, river valleys, rugged mountains and great seas of the polar regions. Thanks to the spacecraft's flexibility, and the ingenuity of flight engineers, the mission also became an observatory of Enceladus, carrying out many close flybys of the exciting little ice moon.

Fig. 3.13 Illustration of the Huygens probe at its final resting place on the surface of Titan. Note the splash mark on the surface behind it. Accelerometer data indicates that the probe bounced and slid on the moist sand (Painting by the author, used with permission from Astronomy magazine)

TOMORROW'S AGENDA

So many revelations have come at the hands of Galileo and Cassini that the gas giants are vying for more missions, but Uranus and Neptune also beckon. If Heidi Hammel were in charge and we had the needed technologies, she says, she would go back to either Uranus or Neptune. "We've done an orbiter at Jupiter. We've done an orbiter at Saturn. We've learned huge amounts about those systems, and they are incredibly rich. The ice giants will be equally rich and unique. We will learn things that we haven't even thought of yet. Many of us want to go to Neptune for a couple of reasons. One is that we're guaranteed an active atmosphere. But two, the moon Triton is hugely compelling."

Carolyn Porco, Principal Investigator of Cassini's imaging team, agrees.

> I think the outer Solar System has really gone under-sampled. The missions are expensive because the targets are far away and they take so long to conduct. There's great interest in understanding the terrestrial planets because we live on one. But if you want to understand cosmic processes, you don't study the inner Solar System; you study the outer Solar System. There are some stars that are like Neptune and Uranus in that they have magnetic fields that are not co-aligned with the spin, and they're even off-centered from the center of the planet. Neptune's, for example, is a third of the way [out]. That's peculiar. It's something about the way they are formed. There's just a lot to learn out there. So if I were queen of the universe, I would put Cassini-type orbiters around Uranus and Neptune.

In fact, several years ago the Planetary Decadal Survey rated a mission to Uranus as a high priority.

However, the prospects of life, or prebiotic conditions, call to other researchers. Some designers envision a Europa "Clipper," a spacecraft that would orbit Jupiter in such a way as to regularly map Europa, studying its general interior as well as its surface characteristics in a series of flybys. Advanced probes might even be tasked to land and drill into the ice, searching for signs of life.

From the time of the Voyagers, Europa has taken the lead as favorite among many astrobiologists. Getting a "cryobot," a submarine probe, into that ocean is a tempting prospect. But the moon presents major challenges. Any life-hosting ocean is estimated to be up to 100 km below a thick ice crust. Getting through this crust is daunting. Experiments have been carried out on frozen lakes in Antarctica, using mockups of probes with heating elements to melt through the ice, but communication cables, which might need to be hundreds of meters to kilometers long, get hung up in the ice as it freezes above. Another cryobot design from Honeybee Robotics[4] proposes a "dangling drill," an active burrowing device suspended from a fine tether that would claw its way through the crust to the ocean below. ESA envisions a simpler penetrator that would embed itself into the ice directly from orbit at high speed without the need of a landing system. Ice forced into the front of the probe could be analyzed in a protected

4. See "The Ice of a Different Moon" by Meghan Rosen, *Science News*, May 17, 2014, p. 20.

instrumented compartment farther back. Europa's surface might be scattered with detritus from its oceans below, but Jupiter's radiation would quickly break down evidence of life.

At Saturn, Titan has been the subject of studies for Montgolfier balloons, blimps, drones and boat probes that could float on the methane lakes of the exotic moon. One such mission, called the Titan Mare Explorer (TiME), would drift for weeks or months on the surface of Ligeia Mare, the second-largest of Titan's methane seas. Studies show that natural lake currents would take the probe completely around the sea to sample the environment at many sites.

Another major target of interest is Saturn's moon Enceladus. Cassini directly sampled its geysers during several flybys and found organic material and salts, all suggesting a long-term, active subsurface ocean. Carolyn Porco would choose Enceladus, with its benign radiation environment, over Europa. "You don't have to bunker your spacecraft with 2 feet of lead to protect yourself. Just take a properly equipped spacecraft and test for big organic molecules and for chirality. Chirality is a test for life. Just fly through the plume and pick up both the solids and the vapor." Porco also says that the drilling required on Europa is not an issue on Enceladus. "Whatever it has in its subsurface ocean is there for the asking. It's accessible. All you really have to do is land on the surface, look up and stick your tongue out. Roughly 90 % of the solids go up and come down again."

Despite political travails and funding interruptions, exploration of the outer worlds will continue. The revelations of the past several decades have constituted a dramatic shift in culture, technology and knowledge as a direct result of the discoveries in planetary sciences. John Spencer, veteran planetary scientist at the Southwest Research Institute, observes that, "No previous generation has gotten to do any of this stuff, so we can't complain. We've been exploring the Solar System for 50 years, and that's like from 1480 to 1530. Think of all that happened: the Portuguese first going down the coast of Africa all the way to Coronado coming into New Mexico in a similar timespan. It must have been amazing."

In the 1950s and 1960s, spacecraft with names like Luna, Ranger and Surveyor took humankind's first steps from Earth, ultimately opening up the frontiers of the Moon to the Apollo astronauts. Like the Lunas and Rangers, perhaps Voyager, Galileo and Cassini are lighting a trail through the outer darkness, a trail that will be trod by tomorrow's adventurers.

Fig. 3.14 The European Space Agency is planning a Ganymede orbiter mission called JUpiter Icy moons Explorer (JUICE). It will launch in 2022, with arrival at Jupiter in 2030 for a three year mission. In concert with this ambitious orbiter, Russia's Institute for Space Research (IKI) is studying a possible Ganymede lander. (image courtesy Dr. Konstantin Marchenkov)

*Fig. 3.15   A future blimp probe
plies the skies of the planet-
moon Titan. How soon will
humans follow? (Painting by
the author)*

Part II
Destinations

Fig. 4.1 Skimming over the rings of Saturn, a sightseeing craft gazes at the chevrons on the outer edge of the B ring. Here, gravitational disturbances from small moons trigger waves of material some 2.5 km high. In the background are (left to right) Mimas, Tethys, and Enceladus, with a tiny moon embedded in the rings just left of center (Art © Michael Carroll)

Chapter 4
The Gas and Ice Giants

The French Impressionist artist Pierre-Auguste Renoir liked to paint outside, *en plein air*. The experience brought a freshness, an immediacy, to his canvasses. No matter how often he painted in the gardens of Giverny or in the forests of Fontainebleau, nature always provided surprises. In his diaries he commented, "You come to nature with your theories, and she knocks them all flat." He could have been talking about planetary science.

The arrival of the Space Age knocked many theories flat. Before it, observers struggled to decipher what they saw across the great distances and under the low light levels of the outer Solar System. Even as late as the 1960s, telescopes revealed little of the mysterious giants. Astronomers had known of Jupiter's Great Red Spot and contrasting belts for centuries. An amateur Scottish observer, William Thompson Hay, had discovered a great white spot on Saturn, providing a clue that the golden planet shared similar weather patterns to Jupiter. Uranus and Neptune stubbornly refused to give up many secrets, as Patrick Moore wrote in his 1961 summary of current knowledge:

> Not much is known about the two outer giants, Uranus and Neptune, which are almost perfect twins. They seem to be made on the same pattern as Jupiter and Saturn, but surface details are hard to make out simply because they are so far away…both the outer planets have satellites. Uranus has five, Neptune two. Of Neptune's attendants, one (Triton) is over 3,000 miles across; the other (Nereid) is a dwarf, and has a very eccentric orbit, so that its distance from Neptune varies between 1 and 6 million miles.[1]

There was so much to learn, but one thing was clear early on. The atmospheres of the giant worlds appeared to consist mostly of hydrogen and helium. Although scientists couldn't go visit the gas and ice giants, they could learn some lessons in the laboratory. Stanley Miller's pioneering work with atmospheres similar to those of the giants yielded organic material and complex hydrocarbons, the building blocks of life.[2]

1. *The Picture History of Astronomy* by Patrick Moore (Grosset & Dunlap, 1961).

2. At the time, scientists believed that the primordial Earth harbored a reducing atmosphere similar to that of the gas giants today. Although this assumption is no longer held, the experimental work is still applicable to outer planet analogies. Miller's later work used different combinations of gases more likely to have occurred on the primordial Earth.

M. Carroll, *Living Among Giants: Exploring and Settling the Outer Solar System*, DOI 10.1007/978-3-319-10674-8_4, © Springer International Publishing Switzerland 2015

In 1953, Miller and his University of Chicago professor, Harold Urey,[3] injected steam into a mix of hydrogen, methane, and ammonia. They subjected the primordial brew to electrical discharges, resulting in a rain of dark organic residue. The materials making up the brownish sludge that Miller found in the bottom of his laboratory jars are called tholins. The possibilities were not lost upon the authors of *Life* magazine's 1966 book *Planets*[4]:

> Incapable of supporting any form of life: that was once the undisputed verdict on the four outermost planets – Saturn, Neptune, Uranus and Pluto…But this verdict is now open to reasonable doubt. The fact that Saturn, Uranus and Neptune have atmospheres at all suggests the presence of life's precursors…recent discoveries have shown that [their atmospheres] are chemically suitable for reactions that may be sufficient to begin life. Also, these distant planets are definitely not, as was once believed, uniformly subject to temperatures of −270 ° F. and below. The wan heat of the sun combined with the internal warmth emitted by the planets have been permanently trapped beneath their atmospheric envelopes.

Astronomer Carl Sagan and atmospheric scientists such as Jonathan Lunine championed the idea that, in similar fashion to the giant planets, Saturn's oversized moon Titan might generate tholins as a result of sunlight reacting with methane. Veteran planetary scientist Ben Clark adds, "In the lab, they try to simulate the atmospheric conditions of Titan with the gases they think are there, and then try to see what kind of particles are there. This work tends to generate these tholins, much like was being done by Carl Sagan and Stan Miller. The stuff's a little dangerous to work with. It's carcinogenic, because it's a mix of all these oddball chemicals."

We now know that this process is common in the upper atmospheres of the outer Solar System, occurring on all the gas and ice giants. It would seem that the chemical foundations of life are scattered through the clouds of Jupiter, Saturn, Uranus, Neptune and Titan.[5]

3. They conducted the experiment subsequently at the University of California at San Diego, where it was published in 1954.

4. *Planets* by Carl Sagan, Jonathan Norton Leonard and the Editors of LIFE (LIFE Science Library, 1966).

5. Triton's atmosphere also generates hydrocarbons, but with far less atmosphere, and fewer chemical reactions take place in its environment.

Fig. 4.2 The apparatus used for the Miller-Urey experiment (Photo montage by the author)

THE MODERN VIEW

With the advent of modern space exploration and advances in remote observation, we have come to know the outer planets as worlds of cloud and mist, where ferocious winds tear towering thunderstorms into banners a thousand miles long. Cyclones as large as Earth rage for years or decades, combining, changing color and disappearing into the multicolored fog. The gas worlds play host to a mix of hydrogen, helium and ammonia, while more methane seasons the bluer ice giants. On Earth, we use ammonia to clean our bathrooms and methane to heat our homes. But the gas and ice giants are worlds of numbing cold, where methane becomes a rain of cryogenic liquid and ammonia falls as snow through poisoned cloudscapes.

Rain in the outer worlds has no place to land. Unlike the terrestrial planets, their surfaces are not ice and rock but amorphous gases and clouds. On all the gas and ice giants, there is no abrupt delineation between solid and gas, no place to stand on these cosmic colossi. Gas giant cores are far more dense than Earth's core, but as one moves out from the center of each planet, the environment transitions from a rocky and metallic solid to a sea of liquid metallic hydrogen, and then to a gas. On Uranus and Neptune, the core pressures are not high enough to force hydrogen into a liquid metal form.

The scale of the gas giants dwarfs anything in the terrestrial neighborhood we call home. In essence, our Solar System consists of the Sun, Jupiter, and a few incidental bits. Jupiter itself is more massive than all the other planets and their moons combined. Over 1,300 Earths would fit inside of Jupiter. Saturn is not much smaller, and its ring system would stretch two-thirds the distance from Earth to our Moon. Uranus and Neptune are twins in size, both encompassing as much volume as 60 Earths. The meteorology of Uranus is not as subtle as once thought, and its weather joins Neptune's as a machine, generating great methane rainstorms and cyclones the size of the terrestrial worlds.

All of the giant planets spin rapidly. Jupiter's day lasts a scant 9 h 55 min. Saturn rotates in about 10½ h. Because the planet is far less dense than Jupiter, its spin flattens its poles substantially. Uranus and Neptune turn once each 17¼ h and 16 h, respectively.

Jupiter, Saturn, Uranus and Neptune have reducing atmospheres, which are atmospheres dominated by hydrogen and helium, the ancient building blocks of the Solar System. The ratios of helium and hydrogen within both Jupiter and Saturn are similar to the composition of the Sun itself. Uranus and Neptune have evolved slightly different blends of gases, more abundant in methane. While the outer two blue worlds have at least 15 % helium in the air, something has drained the helium from the upper atmospheres of Jupiter and Saturn. It may be that the helium in the atmosphere of the gas giants has ended up in the super dense "seas" of liquid metallic hydrogen near their larger cores.

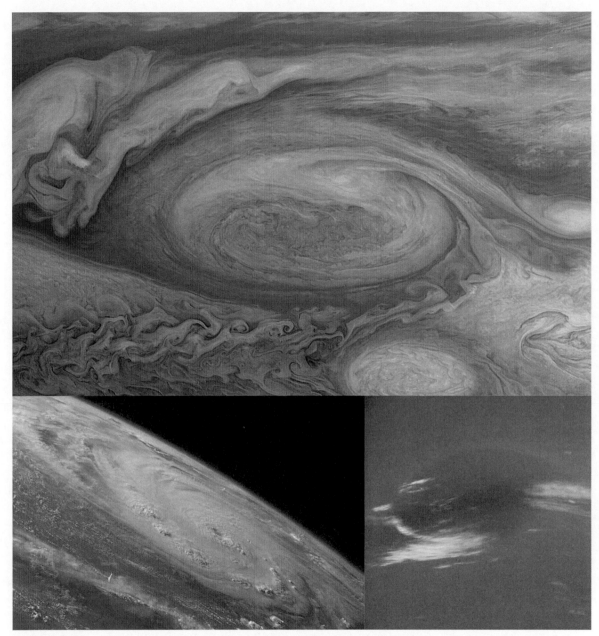

Fig. 4.3 Clockwise from top: Jupiter's Great Red Spot, Neptune's Great Dark Spot, and Hurricane Felix compared (not to scale). Hurricanes on Earth last for weeks, while the Great Dark Spot lingered on a scale of months. Jupiter's massive cyclone has persisted for centuries (Photos courtesy (top) of NASA/JPL. Processing by Bjorn Jonsson; (lower left) NASA/JPL; (lower right) NASA/JSC)

Taken as a foursome, the giants share similar patterns as well as basic differences. Cloudy belts encircle all four worlds parallel to their equators. The cloud bands are wracked by incredibly strong winds, but they remain stable within their latitudinal paths. On Earth, continents divert air currents, while storms continually break up the bands of clouds. Terrestrial storms come and go on the order of days or, in the case of hurricanes and monsoons, weeks. But the bands and giant storms of the gas giants may last for decades or even centuries. Astronomers have observed Jupiter's Great Red Spot (GRS) for nearly 400 years. Initially believing it to be an enormous volcanic eruption, they eventually tumbled to the fact that giant worlds could generate giant cyclones. Two Earths would fit inside Jupiter's burgundy hurricane with room to spare.

The cloud banding on the gas giants is most obvious on Jupiter and Saturn, where the Sun shines most directly on the equator. Despite its distance from the Sun, Neptune's bands are nearly as well defined. Even Uranus – whose axis is tilted so that the planet essentially rolls around the Sun on its side – has subtle banding. Many planetary meteorologists predicted that the weather of planets far from the Sun would be comparatively calm. With less heat to drive air movements, they assumed that Saturn's winds would be more subdued than Jupiter's, and that Uranus and Neptune would continue the trend toward quieter skies. But distance does not bring calm, says Caltech's Andy Ingersoll, the world's leading expert, Principal Investigator for planetary atmosphere studies for Voyager. "The idea we get is that the winds don't decrease as you move out in the Solar System. That was a significant finding. It was borne out as Voyager moved on to Uranus and Neptune." Most dramatic in the trend was Neptune, because the distant planet receives only 5 % of the amount of solar energy that Jupiter does. In fact all the giant outer planets have stronger winds than Earth does. "Why should it be that the winds are stronger in the outer Solar System?" Ingersoll asks. "Even among the giant planets, why should they be stronger at Neptune than at Jupiter?"

Eastward and westward winds segregate the outer planet clouds into patterns of dark belts and light zones. Earth has something akin to these in each hemisphere. The trade winds drive a westward air current at latitudes near the equator, while the jet stream drifts along in eastward currents at mid-latitudes. Neptune's belts are arranged similarly, with two major currents in each hemisphere. Voyager saw only one hemisphere of Uranus, but ground-based data reinforces the suggestion that similar bands play across its globe. Jupiter has five or six belts and the same number of zones in each hemisphere. Zones tend to be sinking air masses, while the belts contain ascending air. Saturn's belts are harder to see, lurking beneath golden hazes, but the ringed giant has a similar number of belts with much higher winds than those found on Jupiter.

Observations and computer models indicate that there are three layers of clouds on each of the gas giants. On Jupiter and Saturn, the highest deck consists of ammonia. Below this floats a brownish mixture of ice crystals, part ammonia and part hydrogen sulfide. Underneath this, mid-deck, mists of water-ice or water vapor simmer in the depths. Neptune and Uranus also appear to have their cloud decks arranged similarly, but their colder temperatures do not form high ammonia clouds. Instead, brilliant white clouds of methane float above a deeper blue layer. The lower cloud blanket may also be methane. Still farther down, water clouds may contain ammonia, but if so, they are hidden beneath the middle deck.

A host of processes spawn the rich brew of complex chemistry within the skies of the outer planets, including rising currents from interior heat, cloud formation (condensation) and photo-dissociation, the process in which sunlight splits molecules of gas. As a result of these global chemistry experiments, the clouds of Jupiter and Saturn are painted in rich oranges, tans, browns and blues. Farther out, methane tints Uranus and Neptune toward the blue end of the spectrum. Neptune's clear air reveals a rich teal cloud deck, while hydrocarbon hazes tint Uranus to a pale shade of blue-green.

One feature that the gas and ice giants share in common is that each is encircled by a ring system. The rings vary in scale, completeness and brightness, ranging from Saturn's spectacular system to Neptune's incomplete ring "arcs." The difference in rings may be due, in part, to age. Saturn's extensive ring system may have formed most recently. The genesis of ring systems has been the subject of spirited debate. Did the rings come from the destruction of a moon? Did a wandering rubble-pile asteroid make its way too close and get pulled apart by the planet's gravitational forces? Are the rings merely the remnants of a moon that never formed in the first place?

THE KING OF WORLDS

After decades of space exploration and centuries of observation, investigators have been able to piece together a clear picture of the skies of Jupiter. In the stratosphere, that part of the sky just above most of the weather, the air is clear. It is laced with poisons of hydrocarbon hazes, refracting sunlight toward the blue due to Rayleigh scattering, just as air does in Earth's skies. Just below this layer float delicate white wisps of ammonia ice crystals. The patchy cloud deck rides updrafts and flattens out at the boundary between stratosphere, above, and troposphere. Beneath, pressures increase and temperatures become warmer. At the top of the bright ammonia clouds, the pressure is about one-tenth that of Earth at sea level – 0.1 bar – and the temperature is −152 °C, on the rise. Ammonia snows fall from the clouds, headed for the depths.

Next, the rich rust-brown cloud deck spreads its cottony billows, giving Jupiter its dark banding. This layer of ammonium hydrosulfide cloud, with a base about 80 kn below the ammonia deck, may be laced with rich organic compounds generated from a number of forces: heat and radiation flowing from within Jupiter, as well as complex chains of amino acids cooked from the clouds by powerful lightning. Compared to Earthly skies, lightning is fairly rare here, but a Jovian bolt packs enough punch to energize a small town for days.

Oval storms the size of Earth's continents chew away at the cloudy bands around them. Feathery mists trail for hundreds of miles along powerful jet streams. Blemishes of rich color dance around each other, merge and dissipate in a ceaseless interplay of color, motion and shifting form.

Here and there, pale blue-gray clouds break through the ruddy ammonium hydrosulfide plain, water clouds boiling up from below.

The water cloud deck is the lowest, floating nearly 100 miles below the highest ammonia cirrus. Here, temperatures rise above the melting point of water, allowing water vapor to billow into clouds. Under this cloud layer, in an eternal night of crushing pressures and searing temperatures, the hydrogen that makes up most of the air compresses into a liquid, and then into a bizarre, electrically conductive fluid metal.

Jupiter's belts and zones may be more than skin deep. Atmosphere acts as fluid. Fluids in a rotating sphere line up with the axis of rotation. The interiors of all the giants may be arranged as a series of stacked disks, each rotating at its own speed. Zones may simply be the surface effect of these rotating cylinders.

Jupiter is encircled by fine rings of dust the consistency of smoke. These rings are so thin that they were undetected until the Voyager spacecraft looked toward Jupiter after its flyby and saw the rings illuminated from behind.

The greatest challenges offered by Jupiter to future explorers are its immense gravitational field and its deadly radiation. Its molten core is the size of a terrestrial planet, and it generates magnetic fields millions of miles into the surrounding space. Even before the Pioneers and Voyagers demonstrated just how deadly those energetic fields were (see Chap. 3), observers knew something odd was going on. Powerful radio beams emanated from the Jovian environment in sync with Jupiter's rotation, while others seemed to follow the orbital motion of Jupiter's moon Io. It was all very

Fig. 4.4 Jupiter's belts and zones are seen from polar perspectives in these Pioneer 11 images, processed by Ted Stryk (Image courtesy of NASA/JPL/Ted Stryk)

mysterious to Earthbound observers. With in-situ measurements by spacecraft, we now know that Jupiter's magnetic field is 20,000 times as powerful as that of Earth's. Electrons and other lethal charged particles are entrapped within a bubble surrounding Jupiter, extending out some 3 million km into space around the planet. The solar wind drags a tail away from Jupiter a billion km behind it.

How is it possible to adequately explore such a hostile environment? At Johns Hopkins University, the Applied Physics Lab's Ralph McNutt has been giving the problem some thought. "If we suppose the sky is literally the limit, with technology and budget, then you could think of getting some sort of an ultrasonic aircraft into the upper fringes of the Jovian atmosphere. Jupiter has one heck of a gravity field, so by the time you slow something down enough to get down to the nominal visible surface, you've already accelerated up to tens of kilometers per second of speed. You're talking about dropping into a gravity field that accelerates you up to about as much speed as the Earth has going around the Sun." At that speed, a hypersonic craft could cover a lot of territory, sailing the alien skies of the most giant of worlds.

SATURN

The magnificent lord of the rings receives just one-fourth of the solar energy that Jupiter does (one hundredth that of Earth). Studies suggested that its meteorology would be less vibrant than Jupiter's because of its reduced solar heating. Through their telescopes, observers could see muted belts and zones, subdued versions of those on Jupiter. But the Voyager fly-bys revealed Saturn's weather to be just as riotous, and with its own unique style. Nearly three decades later, the Cassini Saturn orbiter reinforced the alien nature of the golden world compared to its Jovian sibling. The longevity of its hazes and cloud features is counterintuitive to those who study Earth's mercurial weather. It was a lesson that the outer planets would offer again and again: Earth analogies didn't always work for the giant planets.

A golden fog is responsible for Saturn's glorious tint and subdued features. The haze forms as ultraviolet light from the Sun forms hydrocarbons such as acetylene or ethane. These longer-chain hydrocarbons build up and gradually condense into a haze floating above the belts and zones.

The planet's winds are among the fiercest in the Solar System. Saturn is a world torn by supersonic gales, where yellow-white clouds shear across dark bands like the banners of Olympic racers. When the Cassini orbiter settled in for its long-term reconnaissance, a band of powerful thunderstorms encircled the southern hemisphere, limited to a zone roughly 37° south of the equator. This "Storm Alley" generated a series of long-lived bright clouds laced with electrical discharges.

Lightning is typically associated with water clouds, and the water clouds on Saturn are buried deep. Water first condenses around 20 bars of pressure, some 200 km below the visible cloud tops. Internal heat causes updrafts, bringing the water up to the 10-bar altitude, a vertical trip of about 70 km. At that level, it turns to ice. Lightning on Earth appears to occur at altitudes where liquid water is turning to ice. Particles of ice and liquid water collide, causing a separation of charge. Models predict that lightning on Saturn would begin at about the 10 bar level, about a 100 km down from the cloud tops.

The fact that the thunderstorms were limited to such a focused region baffled researchers. But as studies continued, patterns slowly emerged, and some researchers suspected that the problem was one of time. Earth's atmosphere evolves chaotically on timescales of days to weeks. The gas and ice giants are so much bigger that they may exhibit some of the same aspects of unpredictable behavior, but those may evolve over much longer timescales. Was there, in fact, something special about Storm Alley?

The answer came with the change of seasons. As autumn came to the northern hemisphere, the storms abated in Storm Alley, and one of the most dramatic eruptions of clouds appeared in the north. Water clouds boiled into the upper atmosphere, completely encircling the globe within weeks. And the spectacular storm appeared exactly as far north of the equator as the thunderstorms had been from the equator in the south. Saturn's great rainstorms, it seems, are seasonal.

Saturn's weather offers up another beautiful and tantalizing formation in geometric forms that encircle the poles. In the south, a great vortex stares from concentric cloud circles. The storm's steep rim rises 40–64 miles above the surrounding maelstrom. Despite 550 kmph winds, the

Fig. 4.5 Saturn's "Storm Alley" displays a chain of swirling vortices at 37° south of its equator (left). The great storm of 2011 appeared as spring arrived in the northern hemisphere (Image courtesy of NASA/JPL-Caltech/Space Science Institute)

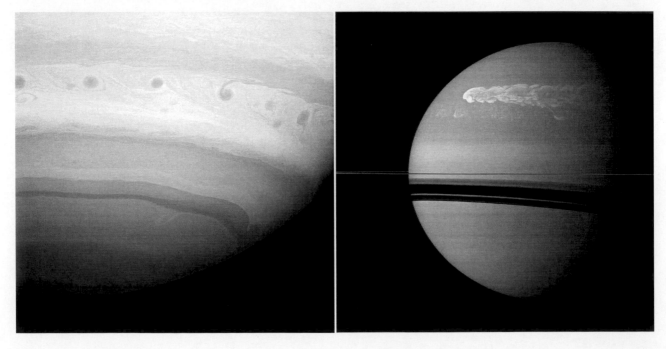

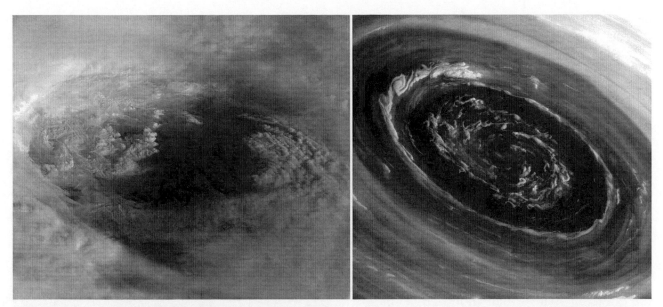

Fig. 4.6 *The eye of Hurricane Isabel (left) would be dwarfed by Saturn's south polar vortex (Left: ISS image courtesy of NASA. Right: Image courtesy of NASA/JPL-Caltech/Space Science Institute)*

Fig. 4.7 *The great hexagon could encircle four Earths. In the above view, we can see the rings beyond the planet at the top of frame. Below is a detail of the elegant cloud forms at one of the corners (Image courtesy of NASA/JPL-Caltech/Space Science Institute)*

storm remains centered directly over the south pole. At Saturn's opposite pole to the north, an enormous hexagon borders territory the diameter of two Earths. Glimpsed by the Voyagers in the 1980s, the bewildering flow of air is stable and long-lived. To some, the Voyager images suggested that the hexagon might be a short-lived disturbance that was being forced by some adjacent vortices or spots. But Cassini images reveal that the hexagon is still there, and it remains remarkably similar in size and shape.

Another unique feature of Saturn's polar regions concerns their color. When Cassini arrived, Saturn was coming out of its northern winter. Ring shadows had darkened the northern hemisphere for years, and clouds there were distinctly bluer than the rest of the planet. Cassini Imaging Team PI Carolyn Porco explains, "The rings cast a shadow that really makes the atmosphere cold. One idea is that perhaps the clouds just sink because the air gets so cold that the level where clouds can form gets lower and lower in the atmosphere. Above the clouds gets clearer and clearer, and you're just getting a lot of Rayleigh." As Saturn passed equinox, shadows shrank toward the equator and the north warmed, and Saturn's golden glow returned there. As the ring shadows shifted to the southern hemisphere, the clouds there are becoming distinctly more blue, a uniquely Saturnian seasonal effect.

The grandest of all the giant planet ring systems are those at Saturn. From the inside to the outside edge of those rings, 67 continental United States would fit end to end. Despite their vast diameter, their thickness is, in most places, equivalent to a three-story building. They appear to be composed primarily of ice, although some portions may contain ice-covered rock. Within the rings lie many small moons that may be remnants of the rings' formative epoch. Astronomers name the rings by letters in the order of their discovery. The brightest central ring owns the label "B"; it was discovered at the same time as rings on either side of it, the "A" and "C" rings. It took a spacecraft flyby to discover the "F" and other faint rings.

Fig. 4.8 *The ring shadows cool an entire hemisphere of the planet, causing the clouds to sink and appear bluer toward the pole. This Cassini view looks down on the southern pole of Saturn as winter encroaches (Image courtesy of NASA/JPL-Caltech/Space Science Institute)*

Fine waves, ridges and gores undulate throughout the system, propagated by the gravity of small moons. Larger moons cause resonances with the particles, clearing gaps in the rings.

Dan Durda, an expert on asteroids and small moons of the Solar System at the Southwest Research Institute, says the rings have much to teach us. "All the little wavelets and spiral density waves, everything we see is Isaac Newton's playground. It's all just gravity. It's just Newtonian gravity between particles, and to have that wide an expression of the myriad phenomena that can be formed from that is just remarkable. The Saturn ring system is probably one of the best natural laboratories for seeing that."

Two such moons that have been imaged at relatively close range are Pan and Atlas, and they both share a very bizarre form. The moons have

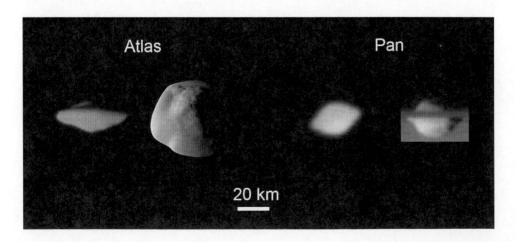

Fig. 4.9 *The small moons Atlas (left) and Pan have remarkable rims at their equators (Image courtesy of NASA/JPL-Caltech/Space Science Institute)*

pronounced ridges extending from their equators. These ridges are thought to be ring material, and this tells us something about their evolution, says Porco:

> You have to have them embedded in a thick disk to begin with. There's a regime where a moon is not big enough to make a big gap, so it's interacting with the rings all the time. I would imagine that particles accrete on to the moon and get kicked off with the same frequency. But imagine the moon is getting bigger and bigger, and the gap around it is getting bigger and bigger, and finally it clears a 360° gap. Once it does that, it's still accreting stuff on it, and it will accrete if the ring system around it has gotten flattened. That's the requirement, that the ring system from which it's drawing material is flatter than the accretion ring itself. The thing is that you just have to have a net accretion on the moon, which I assume happens once the gap appears. The net result is that gravity accretes material around the equator – accumulates it around the equator – and after a while the moon opens a gap so much that you can truncate the accretion.

The process may have stopped long ago, as the moons' orbits today are thought to prevent the material around them from settling onto their surfaces now. This timeline is backed up by other examples in the Solar System, says, Porco. "Look at Iapetus' ridge: It looks old, so these ridges can survive for some time. If the ring arcs at Neptune are residual from an extensive past ring system, we may see the same kind of equatorial bulge on some moons out there."

Tilmann Denk, Cassini imaging team member at the Free University of Berlin, thinks of Saturn's ring moons in another context: "I often wonder, if humankind would travel to Saturn, what would they do there? In public talks, I sometimes mention that building hotels at the north and south poles of the ring moons Atlas and Pan might be the most terrific places to spend vacations in the Saturn system. You would just be ~10–20 km above the ring plane, and the viewing of this ultra-huge disk would change within hours (for 'diurnal reasons') and within travel time from the one hotel to the other, so you could have sights on the lit or unlit side of the rings."

Denk also points out that the outer irregular satellites would provide sites as outposts. They are "easy to reach, with a good view of the Saturn system, but no need for permanent attitude control. From there, the 'daily explorations' might be started and coordinated."

However, some locations in the rings may still be forming new moons. One such region, spotted by planetary scientist Carl Murray[6] in April 2013, appears as a distinct brightening more than 1,200 km long. Murray and other researchers estimate that the moon is less than a kilometer across, invisible to Cassini's robot eyes until something hit it, causing the ring material around it to flare up. The moon may have formed quite recently, within tens of years or a few million, but its fate is unknown. For the clump of material to survive as its own natural satellite, it must remain stable long enough to migrate away from the ring into a more

6. Dr. Murray is at Queen Mary, University of London.

clear orbit of its own. Scientists will be watching and have targeted Cassini for closer encounters in the future. For the first time, they may be witnessing a scene that has played out, over and over again, across 4 billion years of Solar System history – the birth of a new moon.

URANUS

It's been called "the most boring planet in the Solar System." To the human eye, its blue-green disk is nearly featureless. But the smallest of the giants is far more dynamic than our first reconnaissance implied.

Fig. 4.10 Birth of a moon? A flare of ring particles at the edge of Saturn's A ring may signal the formation of a new moon as it accretes and migrates away from the planet (Image courtesy of NASA/JPL-Caltech, Space Science Institute)

Uranus' remarkable color stems from the methane in its atmosphere. Methane absorbs red light, leaving the bluer parts of the spectrum to reflect back at the observer. Uranus has a calm, transparent atmosphere to great depth. A ruddy haze shifts the color of the cloud deck toward the green.

Uranus displays a subdued version of the belts and zones common to its gas giant siblings. During the Voyager encounter of the great green world, the Sun was almost directly over the south pole. Nevertheless, the cloud patterns remained arranged in a similar fashion to the other giant planets, with belts parallel to the equator. Uranus behaved as if it were simply a smaller version of Jupiter, tipped over. This shows researchers that the Sun does not control the orientation of outer planet weather systems. Instead, it is the rotation axis of the planet itself that determines the complicated arrangement of the clouds.

Although belts, zones, vortices and cyclones are clearly seen on the other three gas giants, the weather on Uranus is far more muted. What sets it apart? Jupiter, Saturn and Neptune all put out more energy than they receive from solar heating. Their internal heat drives the weather. But Uranus is far colder compared to its surroundings. The temperatures on Uranus and Neptune are nearly equal at −161 ° C, even though Neptune receives only 4/9 the solar energy that Uranus does. In fact, Uranus' temperature appears to be in equilibrium with incoming solar energy, leading to an atmosphere that is less mixed from interior to surface.

Many researchers attribute the low Uranus heat flow to whatever caused the planet to spin on its side. Its odd orientation may be due to a colossal pileup with a roaming planet. Such an impact would be enough to disrupt any internal heat source, giving it an opportunity to radiate far more effectively in a short burst of post-collision activity. Some modelers suggest that the heat source may have shut down completely. But although an impact seems likely, given Uranus' wildly tilted axis, theorists have put forth other reasons for the planet's lack of heat flow. One idea, for example, has to do with convection. The icy mantle in Uranus may be layered in

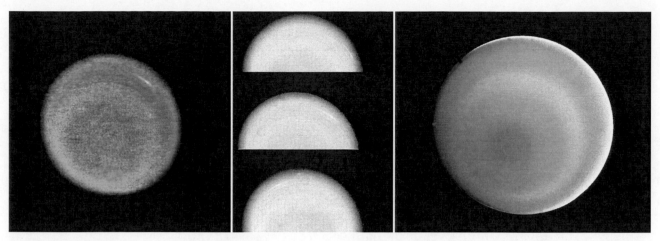

Fig. 4.11 Even the best
Voyager images of Uranus
showed few cloud features
(Image courtesy of NASA/JPL)

such a way that it prohibits convection from occurring. The core produces heat, but the heat is simply not escaping.

During Voyager's encounter with the Uranian system, a polar hood of dark haze spread across the south (the illuminated pole). A cold region stretched from about 10° to 40° latitude. A bright band, concentric to the pole, formed a loop down at the 50° latitude. The band may have consisted of methane clouds and upwelling hazes. Dark bands bracketed the bright one at 20° and 65° latitude. These regions reveal deeper, descending air masses. Unlike Jupiter, where bright clouds are the coldest, Uranus' bright band was warmer than the dark areas. The difference may result from the variation in latent heat on Uranus' methane and Jupiter's ammonia clouds.

The cloud deck on Uranus is buried deep in the atmosphere, beginning at the .9 bar pressure level. The upper atmosphere consists of molecular hydrogen, some atomic hydrogen, helium and haze layers of hydrocarbons. Scientists monitored Voyager's radio waves while the spacecraft passed behind the planet, and were able to chart lower structures. The base of the cloud deck seems to rest at the 1.3 bar level, where temperatures are about −192 ° C. Many narrow cloud bands encircle the planet near the equator. The few discrete clouds that Voyager spotted yielded equatorial wind velocities of up to 580 kmph. The visible clouds are crystals of methane ice and may be welling up from below.

Voyager left us with a portrait of a cold, distant world simmering in quiescence. "People pretty much ignored Uranus after that, because why bother?" says, Heidi Hammel, who carries out research at several major telescope facilities around the globe, as well as through the Hubble Space Telescope:

If Voyager couldn't see anything, surely we wouldn't see anything with ground-based telescopes. Hubble was launched in 1990, just four years after the Voyager flyby, but nobody wanted to look at Uranus; that's a loser proposal. But then I was at a DPS [Division of Planetary Sciences] meeting in 1994, and there was a poster about using Hubble to find moons around Uranus. On the poster was this picture of a planet with moons marked around it, and I said to the author, Ben Zelner,

"What's this picture in the middle?" He said, "We were doing deep imaging with Hubble to look for moons, so the images were overexposed for Uranus, but each pass was a series of three images, so we took one underexposed of Uranus just so we'd have a picture to put on our poster." And this picture showed features. There was a belt and discreet clouds, very bright, and I said, 'Well, that's not what Uranus looks like.' And he said, 'That's what it looked like to us.'"

Hammel obtained the other two pictures and was able to measure a new rotation period for the planet. Uranus was now past solstice, so more of the planet was illuminated and it was at more of an angle to the Sun than during the Voyager encounter.

In the interim, Erik Cocoshka followed up the earlier work using the Hubble Space Telescope to image Uranus. He took what Hammel describes as "amazing" pictures of Uranus, where the entire planet was speckled with atmospheric features.

For Hammel, that was when a transition began. "A lot of us started looking at it with the HST. Things really got exciting when I started collaborating with Imke de Pater and we started using the Keck telescope. I put in a proposal to do a series of images so we could track the winds. These clouds were appearing in the northern hemisphere, the hemisphere that was finally getting illuminated after being in darkness for twenty years. It turned out that Imke had been using the Keck Observatory to do ground-based imaging in times that were overlapping with my Hubble imaging. We combined forces and got very good datasets by combining these datasets. That started our long collaboration."

About 4 years before Uranus reached equinox – the point at which the entire planet would be illuminated from pole to pole – the observers started seeing very bright features. Hammel and others suspect these may be methane cumulous clouds punching up through the cloud deck and then subsiding again. By 2007, astronomers were tracking dozens of clouds. "In

Fig. 4.12 The new Uranus. Clouds and defined bands have formed since Uranus approached its equinox in 2007 (Image courtesy Lawrence Sromovsky, Pat Fry, Heidi Hammel, Imke de Pater. Used with permission)

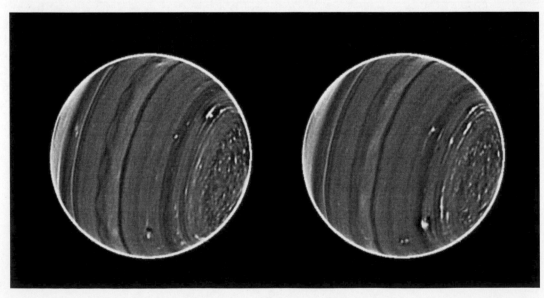

one picture you can count three times as many features as Voyager saw in total." In some events, dark storms would lap each other around the planet, merging into a larger storm that would brighten. Hubble images revealed a dark spot similar to Jupiter's Great Red Spot.

Uranus is clearly not the same planet that Voyager saw nearly three decades ago. How long will it continue to be this active? Hammel isn't sure. "It's an 84-year period. From Voyager in 1986, we did a quarter of the way around the Sun to the 2007 equinox. It's going to be more than a decade until we get to the next solstice. Could it be that this is all happening in equinox and by the next solstice, that hemisphere that has been so dynamic and changeable may be all blank again? We're watching these changes right now. Uranus is so dramatic in its changes."

NEPTUNE

In 1989, after twelve long years in space and several near-disasters (see Chap. 3) *Voyager 2* reached the Stygian environs of Neptune. Neptune's rich sapphire face stands in stark contrast to the soft green of Uranus. Its clouds are intrinsically bluer than those of Uranus. It seems that they contain some kind of coloring agent in the atmosphere that stains them. The reason for Neptune's remarkable blue is the same as that for Uranus' green – missing red light. Three percent of the atmosphere weighs in as methane, and methane molecules are incredibly efficient at absorbing certain wavelengths of light in the red.

As the blue behemoth grew in Voyager's field of view, its imaging system resolved amorphous blobs into dark belts and glowing clouds. Even in the earliest of images, it became obvious that Neptune harbored far more active weather than Uranus had just 4 years before. Voyager monitored Neptune's unique ring arcs and, in the final days of encounter, mapped its moons. One of Voyager's most significant atmospheric discoveries was that – like Jupiter and Saturn – Neptune puts out more heat than it receives. The internal heat makes for the dynamic activity in Neptune's skies. White clouds of methane crystals skitter across subtle belts and zones, sometimes spiraling into cyclonic features.

Unlike the clouds on Jupiter and Saturn, Neptune's clouds were difficult to track. The bright discrete clouds on Neptune evolved fairly rapidly, so that clouds seen on one night could not easily be identified on the next rotation. Neptune's weather systems are far less organized than those on stately Jupiter and Saturn. The zones seem to drift and cross from one latitude to another, capped by wispy clouds that come and go randomly.

Like the other gas giants, Neptune's atmosphere is dominated by hydrogen, the most abundant element in the universe. The second most abundant gas there is helium. The third gas, methane, lends Neptune its

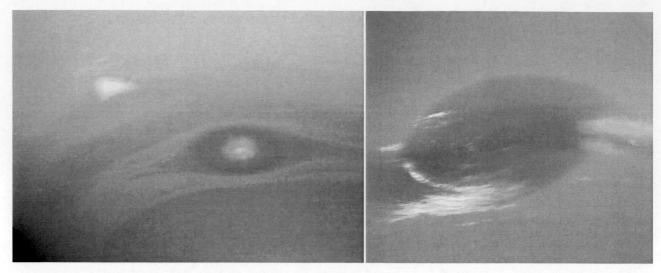

Fig. 4.13 Cloud formations. The bright cloud to the upper left is Scooter. To its lower right is the storm known as the Great Dark Spot 2, or the Small Dark Spot. Neptune's Great Dark Spot was roughly the size of Earth (Images courtesy of NASA/JPL)

distinctive blue. Deeper down, below the blue cloud deck, the atmosphere is likely infused with water and ammonia.

In Neptune's clear upper atmosphere, temperatures hover at about $-180\,^{\circ}$ C. Temperatures rise with depth. At a pressure comparable to Earth at sea level, temperatures reach $-167\,^{\circ}$ C. Beneath this altitude, the air is warm enough for methane to exist as a liquid or vapor. The methane ice crystal cloud deck has its base at a pressure of about 1 bar, where methane can cool and condense into clouds. The clouds billow up above this altitude, bringing methane into the stratosphere. But Neptune's stratosphere has more methane than it should be able to hold at its temperature. The vapor is probably being transported upward from the depths in powerful methane-cloud storms. These storms must traverse 50–100 km of altitude.

For the past three decades, scientists theorized that the large amount of stratospheric methane vapor was due to this "convective overshoot": extremely powerful storms billowing up with methane ice particles that then cooled, condensing into ice particles. But recently, in light of new Earth-based observations, researchers have postulated that the pole of Neptune may have an unusual thermal structure, which would allow methane to leak up into the mid-stratosphere levels. This avoids the need for such strong global methane convection, which some scientists doubt. Oddly, Uranus does not show the excess of methane in its upper atmosphere that Neptune does.

Despite the transient nature of its clouds, Neptune does have a few long-lived structures. One of the first features Voyager resolved in far encounter images was a large, blue storm reminiscent of Jupiter's Great Red Spot. The storm was approximately the same size relative to the planet. Scientists quickly named it the "Great Dark Spot" (GDS-89), a nod to Jupiter's Great Red Spot. But as Voyager resolved more detail, the spot took on its own unique character. The Great Dark Spot spanned the distance of Earth's diameter. This oval-shaped storm appeared to rotate or oscillate on an 8-day period, and flowed quite differently than Jupiter's cyclone.

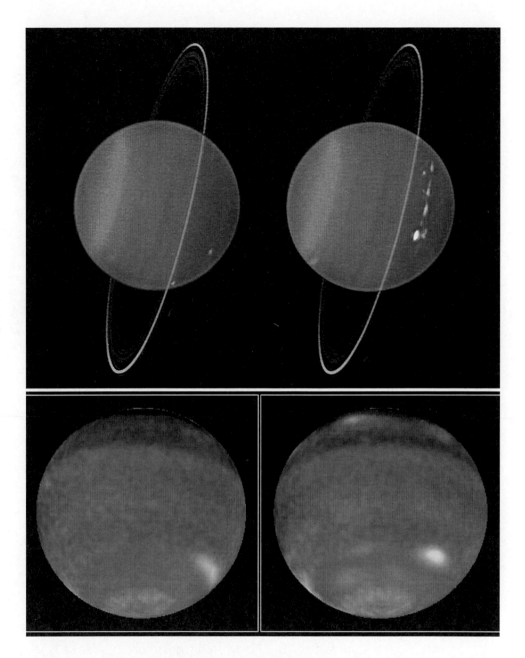

By the time the Hubble Space Telescope came on line in 1993, the Great Dark Spot of 1989 had vanished without a trace. But in 1994, Heidi Hammel spotted a similar storm in HST images of the northern hemisphere. Now known as GDS-94, this storm was slightly smaller. It disappeared within several years and was followed by others. The dark storms appear to have a lifespan of about 5 years.

The white clouds on both Uranus and Neptune usually consist of thin parallel strands. At Neptune, some stretch out in tendrils for hundreds of miles, while smaller versions skate over the deep blue lower cloud deck in hours or days. Voyager team members referred to one such cloud as Scooter. Unlike most of the bright methane clouds, Scooter was rooted deep in the atmosphere. Like the GDS and other cloud features, the cloud moved eastward. It appeared to move quickly compared to the GDS, inspiring its name.

CITIES IN THE CLOUDS – A DIGRESSION

Fig. 4.15 A floating outpost as envisioned by Geoffrey Landis (Painting © Michael Carroll)

In addition to his work on advanced concepts for future space missions at NASA's John Glenn Research Center, Geoffrey Landis is a Hugo and Heinlein award-winning science fiction writer. He recently completed a NASA-funded study to see what it would take to establish a settlement in the clouds of Venus. The floating outpost would cruise at an altitude of about 50 km, where air pressure is equivalent to that at Earth's sea level, and the heat drops to room temperature, a comparatively benign environment. Landis points out that Venus is rich in resources. Its dense atmosphere provides protection from cosmic radiation, and solar energy is abundant at the cloud tops. An added benefit of a Venusian location is that because of the density of Venus' CO_2 atmosphere, a breathable atmosphere is actually buoyant, serving as a lifting gas as helium does on Earth. A 400-m-radius balloon – about the size of a small sports arena – can lift 350,000 tons, a mass equivalent to the Empire State Building.

Landis' report concludes, "In short, the atmosphere of Venus is the most earthlike environment in the Solar System. … [I]n the long term, permanent settlements could be made in the form of cities designed to float at about 50 km altitude in the atmosphere of Venus."

What of cloud cities on the gas and ice giants? The atmospheres of Jupiter, Saturn, Uranus and Neptune are dominated by hydrogen and helium, the two lightest gases around. This makes "lighter-than-air" balloons impossible, but Landis says there is a solution. "It's been proposed to float hot-hydrogen balloons – as long as you heat up the gas inside the envelope to hotter than the ambient, it will float. This needs a power source, though."

The fierce radiation environment surrounding Jupiter may not be an issue. Jupiter's radiation drops to low intensities at the level of the Jovian cloud layers, and temperatures are fairly benign as well. Wind shears are terrific, but within a belt or zone, riding the currents, the breezes may be fairly stable. Perhaps the main problem of living in a cloud city on the giant planets is that your home would be permanent. There doesn't seem to be a practical way to return to Earth from those immense gravity wells.

FINAL THOUGHTS

Exploring the skies of the giant planets themselves may not be possible for humans, except in a vicarious way. The APL's Ralph McNutt thinks that a combination of robotics and humans may ultimately be most effective. He envisions a human outpost on Ganymede or Callisto, where a crew remotely pilots drones in the atmosphere of Jupiter. "You're probably

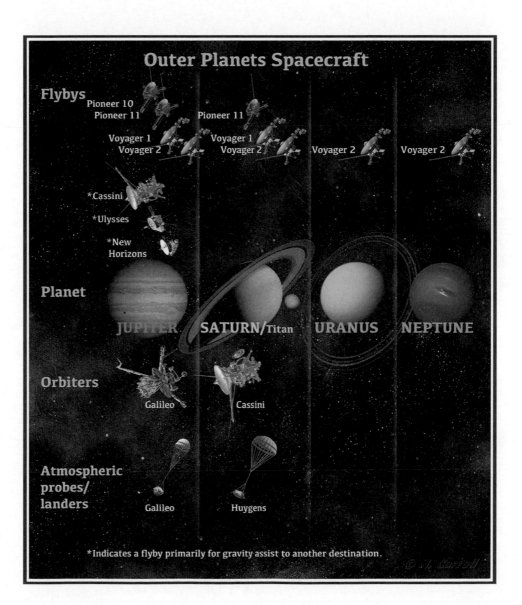

Fig. 4.16 A veritable armada of spacecraft have visited the outer planets and their moons. Notice how the number of missions falls off as the diagram moves toward the ice giants (Image ©Michael Carroll)

talking about some sort of an ultrasonic aircraft, and even at very high speeds, Jupiter is huge, so you're only going to be looking at very small areas. But what we've always found in space exploration is that every time that I take the next step to be able to look at things with another factor of ten resolution, I manage to find so much more and learn so much more. Usually, it includes answering a lot of the previous questions and opening up a whole host of new ones. That's part of the learning experience in all this, in finding out what the world is really like around us."

The famous rocket engineer Krafft Ehricke often gave slide lectures about human space exploration. He would open his talks by saying, "If God had wanted the human race to go into space, he would have put a planet right next door." After the remark, Ericke would show a slide of the Moon. In a similar way, we have been afforded natural outposts next door to the giant worlds: their natural satellites.

Fig. 5.1 The bizarre ice pillars of Jupiter's massive moon Callisto will beckon future travelers (Painting © Michael Carroll)

Chapter 5

The Galilean Moons

German astronomer Simon Maurius was about to see things he had never seen before. They were things that would get him into hot water on an international scale. In the autumn of 1608, Maurius ran across an acquaintance who had a curious tale to tell. John Phillip Fuchs, artillery officer and Lord Privy Councilor to the Margraves of Brandenburg-Ansbach, had just returned from a fair in nearby Frankfurt. At that fair, a Dutch inventor was demonstrating a little tube that made the distant landscape "be seen as though quite near." Fuchs had tried to buy the spyglass, but could not settle on a price.

Fuchs and Maurius attempted to reproduce the telescope using lenses from a pair of spectacles, but could not get the thing to work. A year later, the new inventions were becoming more common – and cheaper – and Maurius was able to buy one. In November or December of 1609, Maurius began to systematically observe Jupiter. He spotted star-like objects moving along with the planet, as he wrote in 1614: "…as Jupiter was then retrograding, and still I saw these stars accompanying him throughout December, I was at first much astonished; but by degrees arrived at the following view, namely, that these stars moved round Jupiter, just as the five solar planets revolve round the Sun."

Maurius began to record his observations on December 29, 1609, in the Julian calendar. This date corresponds to the Gregorian calendar date of January 8, 1610,[1] which happens to be the second night that Italian astronomer Galileo Galilei documented the starry attendants of Jupiter. Galileo published his *Sidereus Nuncius* (Sidereal Messenger) in March of 1610. His paper contained more than seventy of Galileo's drawings of lunar phases and craters, constellations, and – significantly – Jupiter with its moons.

Galileo mentioned the Jovian moons in an earlier letter dated January 7, so he probably spotted the four large moons at least one night before Maurius did. At any rate, Maurius did not publish his findings for several years. When he did, in his 1614 *Mundus Jovialis anno M.D.C. IX Detectus Ope Perspicilli Belgici,*[2] Maurius gave recognition to the Italian astronomer, saying, "The credit, therefore, of the first discovery of these stars in Italy is deservedly assigned to Galileo and remains his." But followers of Maurius took him to mean that Galileo should get credit in Italy, while Maurius did the true discovering in Germany. No one knows how Maurius felt about this, but he had already had an academic conflict with Galileo over observations of a nova several years earlier. He was undoubtedly eager to avoid another such confrontation. Still, the pressure brought to bear by Maurius' supporters compelled Galileo to write a searing rebuttal of Maurius' "claims" to the discovery.

Maurius' 1614 writings are the first to detail the specific movement of the moons, and it was Maurius who first offered the names that are now assigned to them, names of lovers of Jupiter. These were suggested to him by Johannes Kepler in a letter dated October of 1613, and they have stuck.

1. Being Catholic, Galileo used the Gregorian calendar.

2. Or, "The World of Jupiter discovered in the year 1609 by means of a Dutch spy-glass."

M. Carroll, *Living Among Giants: Exploring and Settling the Outer Solar System,*
DOI 10.1007/978-3-319-10674-8_5, © Springer International Publishing Switzerland 2015

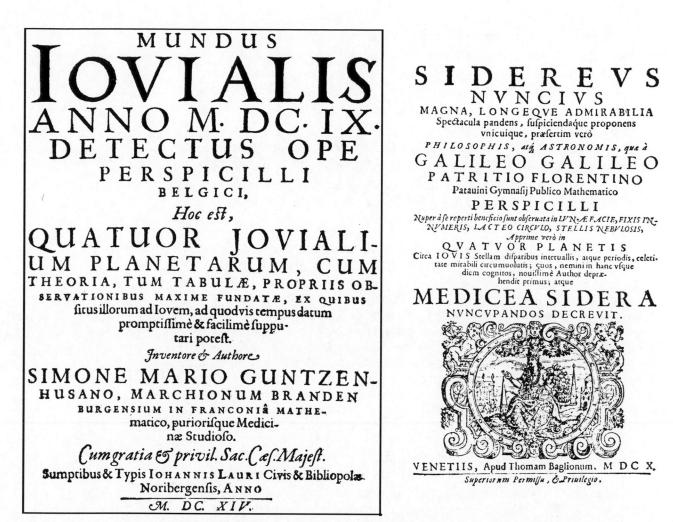

Fig. 5.2 Cover plate for Maurius' Mundus Jovialis (left) and Galileo's Sidereus Nuncius (Both images courtesy Wikipedia Commons. Mundus Jovialis http://de.wikipedia.org/wiki/Simon_Marius#mediaviewer/Datei:Mundus_Iovialis.jpg. Sidereus Nuncius: http://en. wikipedia.org/wiki/Sidereus_Nuncius#mediaviewer/File:Sidereus_Nuncius_1610.Galileo.jpg)

So while the large moons of Jupiter are today referred to as the Galilean satellites, their names come to us from Maurius.

As telescopes grew more powerful, knowledge of the Galileans stalled. The Jovian system is so far away, and the moons so small, that little detail could be made out. They were large enough to be impressive, but frustratingly far enough as to hide their true natures. Over 300 years later, in his 1933 book *Sur Les Autres Mons*,[3] Lucien Rudaux – director of the Meudon Observatory in Paris – wrote, "One sees that the principal satellites are bigger than our Moon and can rival in dimension certain planets. The diameter of Ganymede is almost that of Mars and surpasses that of Mercury, the latter of which is just smaller than [Ganymede]." Rudaux compared the Galileans to the planets neighboring Earth. "You can see small patches of gray which offer evidence of a naturally varied surface structure. It is, therefore, like the Moon and the planets that we can see from Earth.

3. Or… *On the Other Worlds* originally published by Auge, Gillon, Hollier-Larousse (Librarie Larousse, Paris) 1937. Reprinted 1990. In addition to being a fine observer, Rudaux was an accomplished artist. His paintings provide some of the first scientifically-accurate space art in history.

It is possible for us to understand the surface, but what can we know of these sites? Are we seeing accidents of color? Is it flat or rough? Is it sterile or not? In this regard, it is impossible to have a precise idea." Rudaux, being an artist as well as an astronomer, could not resist wondering what these small worlds would be like on a human scale. "It could be possible to set foot on them. But on which should we then walk?"

Over the next 20 years of telescopic observation, astronomers refined the sizes and masses of the four moons, but had made little headway in understanding their natures. Roy Gallant's 1958 book *Exploring the Planets*[4] was able to sum up the current scientific knowledge of the Galileans in a few short paragraphs. The text said, in part:

> Io…is most likely a rocky globe with metals scattered through it. The next large satellite is Europa…an excellent reflector of the Sun's light [that] sometimes appears to wear a dark belt about its equator and to show light polar regions. Generally the satellite appears white… [Ganymede] has dark patches, canal-like markings, and polar caps resembling those on Mars. Like our Moon, Ganymede must be a freezing cold world with an extremely thin atmosphere… Callisto…is a poor reflector of light and appears as a blue-gray with a dark equator… The satellite's dark appearance hints that Callisto is quite different from its companions. Some astronomers regard it as a solid ball of ice. Others say that it is an ice-covered rock core." Well into the 1970s, many researchers thought the surfaces of the Galileans to be cratered and barren….[Ganymede's] terrain is probably very similar to our Moon's…with frozen gases encrusting the rocky plains.[5]

Although the Jovian moons horded their secrets, they did provide some insights into their parent planet. By carefully charting their movements, physicists could determine the masses of the moons to good approximation. The dance of the moons, in turn, revealed the mass of Jupiter itself. Experiments were even done using the Galileans to help determine the speed of light.

In defining the masses of the moons, astronomers attempted to figure out their internal makeup. By the early 1960s, a consensus grew that the Galileans consisted of a mix of water-ice and rock. How that mix was arranged was anyone's guess. Were the moons a jumbled soup of frozen water and rock rubble? Globes of ice with a dense rock/iron core? One thing was clear. Like the planets orbiting around their parent sun, the Galilean moons closest to their parent, Jupiter, were densest, and the farthest one, Callisto, was the least dense. In a sense, the Galileans seemed to be a miniature Solar System.

Beginning in 1964, astronomers began to carefully monitor stars as they disappeared behind the moons in a phenomenon called occultation. If an atmosphere were present, starlight would fade gradually as the star appeared to move behind the moon. But this was not the case; it was becoming evident that the planet-sized moons had little, if any, atmosphere.

4. Exploring the Planets by Roy A. Gallant, Doubleday & Co, 1958.

5. *The Moon and the Planets* by Joseph Sadil; published by Paul Hamlyn, London 1965.

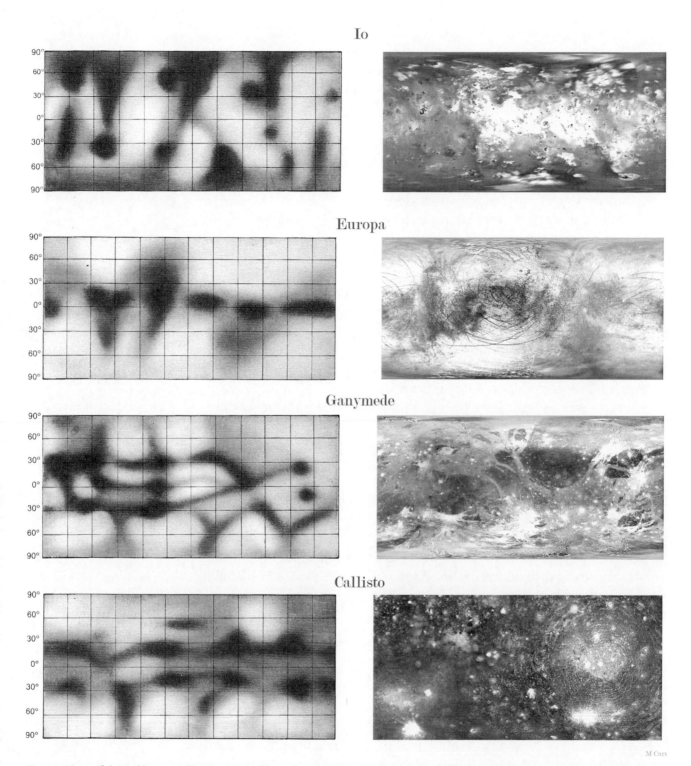

Fig. 5.3 Maps of the Galilean satellites by A. C. Dollfus, using the 60-cm telescope at Pic du Midi observatory, seen next to modern maps of the Galilean satellites (Left: Courtesy of International Planetary Cartography Database. Right: Image courtesy of NASA/USGS)

The astronomers had other tricks up their sleeves. With increased telescopic power, they were able to estimate the temperatures of the satellites. Temperature is an important indicator of surface material. The largest and easiest to study was Ganymede. As Jupiter's shadow fell across the moon, temperature changes told observers that materials on its surface were poor conductors of heat. This, combined with other data, led them to the conclusion that, "the satellite might have a crust of ice covering a satellite-wide ocean of water and ammonia, and below this deep ocean a core of rocks and iron oxide."[6]

Another significant advance in understanding came in 1973, when the *Pioneer 10* spacecraft flew by Jupiter, imaging the planet and its moons. *Pioneer 11* followed a year later. Neither craft carried real cameras (see Chapter Three) but rather had scanning photopolarimeters from which images could be constructed. By today's standards, the images were crude, with resolutions of the Galileans – at best – of about 160 km per pixel. Still, the distant views showed variation in both color and albedo (see Fig. 5.5).

Pioneer 11 had the best view of Io, revealing an orange polar region almost straight on, with a white equatorial band. Io continued to baffle researchers, with its brightly colored surface and compete lack of water in its spectrum. Resolution was 376 km per pixel.

Europa was imaged only once, by *Pioneer 10*. The image has little color variation. A dark region cuts across the center of the half-illuminated globe, just as Rudaux had observed four decades earlier. Europa's highly reflective surface seemed to confirm suspicions that its surface was covered in water-ice.

Two images from the encounters show the tawny face of Ganymede, with nearly uniform color but a dark circular region at the center and a bright polar cap in the north.

Dusky Callisto, the darkest of the Galileans, rounded out the encounter imaging in a distant frame resolving objects 391 km across. Small variations in both color and albedo suggested a lighter region near the equator, but not much else.

The Pioneers also gave flight engineers a refined approximation of densities, but for details, scientists would have to wait another 6 years for the Voyager encounters. The Voyagers fundamentally changed planetary science in many areas. Planetary geophysicist William McKinnon, fellow of the McDonnell Center for the Space Sciences at Washington University, describes the paradigm shift that came about as a result of the Voyager flights. "There have been a vast number of people crawling all over Earth for ages, and we've just started to get to know Earth. Voyager gave us our first look at icy satellites over 35 or so years ago. Simply going out there led us in other directions of thinking."

One of those new directions concerned tidal heating. The small worlds circling Jupiter were expected, by many, to be what Lucien Rudaux suggested: icy versions of Earth's Moon. But interior heating from the gravitational taffy-pull of neighboring moons and the gravity of Jupiter

6. *New Worlds: Discoveries from Our Solar System* by Von Braun and Ordway (Anchor Press 1979).

created a menagerie of geological forms far more active than most predicted. "There had been predictions of tidal heating and what it could do," McKinnon remembers, "but we'd never seen anything like it, and no one had a real appreciation of its power. Even tiny worlds are very active."

The first Voyager close encounter with Jupiter came at 4:42 am PST on March 5, 1979. The air was brisk on this early Pasadena spring morning, but scientists had cloistered themselves in JPL's control room and Von Karman auditorium for hours already. The actual time of closest approach was 4:05 am, but signals from the distant craft took 37 min to traverse the void.

On its way in, *Voyager 1* had snapped several shots of Callisto, Ganymede and Europa. The alluring images showed distinct blemishes, spots and lines, but were so low in resolution (~100 kms per pixel) that no decipherable forms could be seen. But there was more to come. After skimming by the king of worlds, *Voyager 1* carried out its task of surveying the Galilean satellites at closer range. The encounter revealed a miniature planetary system of unique worlds circling the gas giant. Four months later, *Voyager 2* carried out a complementary encounter, studying several moons up close that had been seen from afar by its earlier twin.

Our most detailed insights into the Galilean satellites today came from the Galileo orbital mission 16 years later (see Chapter Three). Rather than a cosmic drive-by, Galileo settled into orbit for a long-term study of the Jovian system, from its magnetosphere to its rings to its satellites. JPL's Robert Pappalardo recounted the foundational shifts in our understanding that resulted from Galileo:

> Ganymede's magnetic field comes immediately to mind. The idea that Ganymede could still be hot enough in its interior to generate its own field, that was pretty huge. Then there was the revelation that the satellites, at least three of them, are fully differentiated [meaning that they had evolved to a point where lighter elements rose to form a crust while the heavier ones settle to the center to congeal into iron cores]. Now we take it for granted, but back then we were asking, "Are they homogeneous, or is there a rocky core?" Europa's ocean was a bit of a pipe dream, but now we have this new paradigm that oceans can be maintained under ice shells. It was thought that as soon as the ice shell was thick enough to convect, that the entire thing would freeze up [from the outer crust to the core], so none of these things could have oceans. That was a big deal.

William McKinnon adds, "It is hard for heat to escape through ice. For a while, we thought that convection, a sort of glacial movement, would get rid of the internal heat. Even though people suspected there might be oceans under Europa or even Enceladus, the data dragged us, kicking and screaming, to the conclusion that oceans are common in the icy moons."

Europa's surface age was another surprise of the Galileo mission results, says Pappalardo. "There was this mottled terrain – the pits – that we saw in the first flyby, that was known from Voyager. We didn't know if it was endogenic or if that was lots and lots of impact craters, in which case the surface was really old. Well, it wasn't impact craters. It wasn't old, and that's a paradigm shift for Europa. Going in, we thought maybe the satellites were a fairly homogenous mix of rock and ice with maybe a small rocky core. Callisto's lack of differentiation in light of the other satellites, which are fully differentiated, is a mystery."

Some of the mysteries left to us by the Galileo mission may be solved soon. Several missions to Jupiter are either under development or underway (see Chapter Three). In the meantime, ground-based research continues to advance.

Even in the largest telescopes, Io subtends only one or two pixels across. Observers must get clever to tease the details from the data, says JPL's Ashley Davies. "You have a couple pixels across Io, so you're seeing the entire satellite at once. Because you have different thermal sources on Io, as Io rotates, some of these sources rotate into view while others rotate off." He continued: "If something big like Loki rotates off, your thermal flux [change in heat] goes down, and as it rotates in again your flux goes up. Even though you're looking at all the energy coming from the entire disk at any given time, you can still detect high temperature events."

A CLOSER LOOK AT IO

With a diameter of 3,636 km, Io is about the size of Earth's Moon, with a surface area equivalent to North America. Io and the other Galileans are tidally locked in their orbits, meaning that the same hemisphere always faces toward Jupiter. Io's day, equivalent to its orbital period, is 43 h.

On the scale of human experience, Io's ochre panoramas must share the grandeur of the great plains of North America or the Russian steppes: undulating golden vistas peppered with soft greens and punctuated, here and there, by the black and orange of sulfurous calderas. And while Earth's northern plains are known for bitter winters, Io's hyperborean climate allows for blankets of sulfur dioxide frosts clinging to its stony, vacuum-clad landscape. But tourists would need to enjoy those landscapes fast. Wearing todays' spacesuits, they would be treated to a fatal dose of Jovian radiation within minutes.

Shakespeare's Juliet famously asked Romeo, "What's in a name?" Jupiter's innermost Galilean satellite is a particularly apt moon to name after a Jovian lover. Jupiter and Io share a uniquely interwoven relationship. Io orbits Jupiter at a distance of 421,800 km, circling deep within the planet's deadly radiation fields. Radiation and electrical currents constantly barrage Io's bizarre landscape, flowing between the moon and Jupiter. The Galileo spacecraft confirmed that Io has a heart of nickel-iron. Its core interacts with Jupiter's magnetosphere, the magnetic field swirling around the planet.

When the Galileo orbiter flew over Io's poles in August and October of 2001, it was searching for a magnetic field emanating from the heart of the moon. It found none. Instead, the craft discovered a tenfold increase in

Fig. 5.4 *Energetic fields and particles tie Jupiter's volcanic moon to the planet. If we could see them, we would notice the flux tube going from the poles of Io to the poles of Jupiter's magnetosphere. At the same time, Io floats inside of a supercharged donut of charged particles, the Io Torus, which encircles Jupiter in a plane inclined slightly from the equator (Painting © Michael Carroll)*

the charged particles that connect Io to Jupiter. Io generates more current than 1,000 commercial nuclear reactors, and much of this flows along Jupiter's magnetic field lines. Galileo traced lines of energy directly above two active volcanic areas. Electrons and ions race along pathways of magnetic field lines, linking those Ionian sources – and others – to Jupiter. This energy superhighway, called the 'flux tube,' channels energy from Io to Jupiter's near-polar regions, powering spectacular aurorae on both the planet and the moon. Io's activity actually beefs up the effects of Jupiter's energetic fields and particles. As planetary geologist William McKinnon puts it, "Jupiter is bad *because* of Io. Io pumps up the magnetosphere with all that sulfur. The satellite systems [of other planets] are smaller and don't affect magnetospheres in the same way."

The radiation bathing Io's surroundings makes for a nightmarish environment, says planetary scientist John Spencer of the Southwest Research Institute. "Io is really stewing in its own juices. It's putting out all this volcanic stuff that is escaping into the Jovian magnetosphere, that is then being accelerated to tremendous energies by the very powerful magnetic fields in Jupiter's system and slamming back into Io, at least into its atmosphere. It's knocking off more stuff in a kind of feedback loop, so the whole inner Jovian magnetosphere gets full of these high energy oxygen/sulfur atoms and electrons that are in lethal doses to any human who would go anywhere near it. They're pretty lethal even to spacecraft; it's hard to build a spacecraft that can survive in that kind of environment."

This high energy sea of electrons causes different areas on Io to glow. Io's sky would be filled with the ghostly shimmer of its sodium cloud. Sodium is very bright, which makes for efficient street lamps. That same yellow-orange streetlamp glow would be visible to the naked eye at night. And there are other auroral glows from oxygen and sulfur and sulfur dioxide that are fainter, but probably still visible with the naked eye. In long exposure images of Io taken while the moon passes through Jupiter's shadow, the glow of volcanoes speckles the surface, telltale signs of feverish heat. The plumes also glow, as does the thin atmosphere surrounding the disk of the small moon. A glow extends above Io's ionosphere where the magnetic field lines from Jupiter connect at points directly facing toward and away from Jupiter. Those clouds of light wobble up and down with every Jupiter rotation, because the planet's magnetic field is tilted. The position of those connection points varies as the magnetic field sweeps by.

The initial discovery of Io's volcanoes came not from observing its surface but rather from observing the stars behind it. It was just 3 days after its closest approach to Io, on March 8, when *Voyager I* took a series of overexposed images of stars for navigation purposes. One of these images caught the sunlit edge, or 'limb,' of a crescent Io. Some 300 km above the surface of the moon hovered what appeared to be an umbrella-shaped patch of fog. A similar feature glowed at the terminator (the boundary between day and night).

Navigator Linda Morabito recognized the ghostly apparition as a cloud. Knowing that Io had no atmosphere, the only logical conclusion left to her was that an incredibly violent volcanic explosion had blasted the umbrella of gas and dust into the sky. Researchers on the imaging team agreed, and infrared sensors confirmed hot spots across the face of the little moon.

This was the first time researchers had been able to see tidal heating in action. The push and pull of Jupiter and the other Galilean satellites causes Io's ground to rise and fall some 45 m each day. This flexing heats the interior of the little moon. All the energy must come out somewhere, and it escapes in the form of the moon's savage eruptions. Those volcanoes explained another mystery, the tenuous sulfur/oxygen atmosphere around the moon. The Voyagers charted 400 active volcanoes. Two decades later, the Galileo spacecraft found more than 120 new sources in just its initial encounter. Geologists estimate that there are roughly 600 active sites, as many as are found on the entire surface of Earth.

"Most people think the story of Io starts with Voyager, but we knew, from ground-based data, that something was up," remembers Torrence Johnson, head of Voyager's imaging team. Johnson and Dennis Matson carried out an extensive set of Earth-based observations of Io. "We saw water absorption features on Ganymede and Europa but none on Io, although Io was just as bright. We saw Io's interaction with Jupiter's magnetosphere, and we saw sodium." Several researchers even predicted volcanic activity there just days before the encounter.[7]

As researchers struggled to understand the peculiar nature of Io, their instruments detected high temperatures on its surface. The numbers were astounding. Io has 100 times as much heat flowing from its surface as Earth does. Roughly 100,000 tons of material erupts from Io each second. Although most of it rains back to the ground in a machine-gun fire of frozen sulfur, some 10 tons escapes every second into the surrounding Jovian magnetosphere. Over 5 % of the Ionian surface is covered by volcanic calderas. "The piece of the puzzle we were missing was the heating process,"

7. Peale, S. J.; Cassen, P.; Reynolds, R. T. "Melting of Io by Tidal Dissipation," *Science,* March 2, 1979.

says Johnson. "Stan Peale suggested tidal heating, but even his team didn't think Io would be cooking enough for active volcanism." Once the scientific community pegged tidal heating as the culprit, planetary scientists rapidly converged on the 'Voyager era paradigm' of tidal heating as a major player in outer Solar System geology.

The temperature of Io's ubiquitous eruptions gives scientists insight into their true nature. Early Voyager data implied low-temperature, sulfur-based volcanism, but Galileo's higher resolution NIMS (Near Infrared Mass Spectrometer) provided more accurate readings. The temperatures seen by Galileo's robot eyes were surprisingly hot. Some people predicted high-temperature sources resulting from silicate eruptions, but most researchers didn't expect temperatures higher than terrestrial eruptions. Early in Earth's history, high-magnesium lava resulted in superheated eruptions in the range of 1,500 °C. But preliminary Galileo data suggested temperatures on Io soaring as high as an astounding 1,700 °C. Current models simply do not explain such temperatures. Johnson adds: "Those temperatures are getting embarrassingly hot, even for rock."

Three types of volcanoes helped shape Io's Dante-esque landscape. The first and most common of the volcanic wounds are lava lakes, collapsed holes in the surface plains. The most classic example of these formations is the feature called Loki Patera (named after the Norse god of fire). Loki is a volcanic crater, or caldera, filled with a lava lake some 200 km across. In the center of its dark, seething plain protrudes a gigantic iceberg of frozen sulfur dioxide. The massive berg is about the size of Africa's Lake Victoria. In similar terrestrial lava lakes, cooled lava forms a crust that breaks up as it collides with the caldera wall. Galileo infrared observations show an incandescent shore along one edge of the lake, similar to the Kilauea and Halemaumau lava lakes on Hawaii.

The second type of volcanic feature common on Io is lava flow. Molten rock has raced across nearly every square meter photographed by spacecraft. The longest flow is Amirani, which stretches across 300 km of rolling plains. A large number of flows are insulated, meaning that a crust of cooled lava covers them. Magma frequently breaks out of the crust many kilometers downstream.

A vast flow issues from the Prometheus volcanic vent, whose eruptions reach 80 km into the void. In the 17 years between Voyager and Galileo, the Prometheus plume had moved some 70 km to the west. "This was really puzzling," says JPL volcanologist Rosaly Lopes. Scientists studying Galileo data realized that the volcanic source of Prometheus had not actually moved, but the lava flow had. "As lava moved across the landscape," she explains, "it crusted over. 70 km away, the lava broke out again, interacting with sulfur dioxide ice." This interaction gave birth to the visible plumes that erupt today.

The third type of volcanic feature is the most dramatic. Pillaen eruptions emit the highest plumes seen anywhere in the Solar System, and the most powerful was found by accident. In December of 2000, Galileo coasted over the Tvashtar Catena region. At that time, a curtain of glowing lava 22 km long fountained from a fracture in the floor of a great caldera. A nearby plume exploded gases over 100 km into the black sky. A 500-km plume was the most powerful yet seen, and was lofted into the air by a volcano called Pillan. Pillaen vents are the source of Io's mysterious high temperatures. The superheated Pillaen eruptions seem to be unique in the Solar System.

One common characteristic of terrestrial volcanoes curiously missing on Io is a shield or mountain built up around volcanic craters. Volcanoes on Io constantly build layer upon layer of new surface, but no tall structures rise from its relatively flat landscapes. It may be that eruptions are so ubiquitous that no one site has a chance to build higher than others. Additionally, most calderas on Io are holes in the ground. They recycle their material inside. Eruptions in a vacuum spread out. Most of the plume material is made up of fine particles or gas, so little buildup occurs at the vent sources.

Fig. 5.5 Sunbeams across a volcanic eruption above Io's limb. The New Horizons grabbed this snapshot of the great plume wafting 330 km over the northern polar regions of Io. The plume comes from Tvashtar (Image courtesy NASA/Johns Hopkins University Applied Physics Laboratory/Southwest Research Institute)

Although the mountains on Io are not typically part of volcanic constructs, they do seem to be related. Many large Ionian mountains are near calderas or vents. This has led some scientists to postulate that Ionian mountain building is quite alien in nature. The mountains appear to form by thrust faulting, and eruptions may break through farther down the line of the same fault. Other fractures may reach deep into the subsurface structure, creating new conduits for lava and gas to escape.

One such site is Tohil Mons, a 6-km-high crescent-shaped ridge bordering a caldera. Numerous landslides scar Tohil's mountain slopes, but the debris seems to have disappeared into the adjacent caldera floor. Apparently, the crater interior was covered by fresh lava flows sometime after the landslides. Tohil seems to have been uplifted along faults adjoining the volcanic crater. Whether the mountain arose after the formation of the caldera, or whether the caldera formed after the mountain was uplifted along a fault, remains unknown.

Another of the unsolved mysteries of Io is its extraordinary coloration (see Fig. 5.6). Most of its brilliant surface hues can be attributed to sulfur dioxide. Laboratory tests show that sulfur dioxide transitions through dramatic color changes as it cools. Molten material is most often black, changing as it cools into red, orange and yellow. Sulfur dioxide frost also powders the landscape in blue and white.

Some areas of Io have been dubbed 'golf courses.' The Ionian golf courses are not as finely tended as the fairways of Pebble Beach, but they are a dramatic green. The green hues of golf course terrain could be caused by sulfur, or by a glassy volcanic glass called olivine. The dynamic geologic forces at work on Io

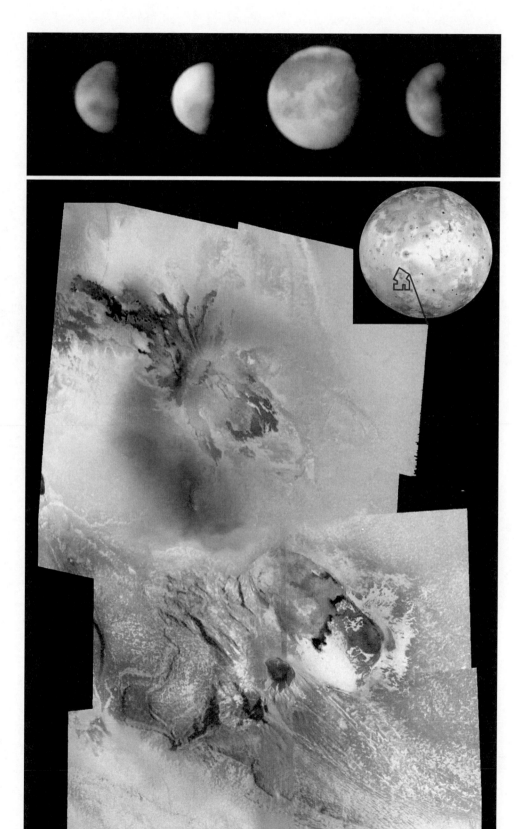

Fig. 5.6 Top: Intriguing views of the Galilean Satellites from Pioneers 10 and 11. Left to right: Io, Europa, Ganymede and Callisto, not to scale (Image processing © Ted Stryk/JPL). Bottom: The active volcano Culann Patera (top center) shows off Io's bright pallet. It lies just north of the summit of Tohil Mons. Both dark red and black lavas flow from Culann, as well as diffuse, inner and outer rings of red and yellowish sulfur particles from explosive plumes. Molten silicate rock mixes with subsurface reservoirs of sulfur and sulfur dioxide to produce the plume deposits. The green color at the center of Culann and inside the older volcano Tohil Patera (center right) may form "golf-course regions" when red sulfur plume deposits land on dark black silicate lava flows and form a green veneer. The small white patches on and near the mountain Tohil Mons might be deposits of sulfur dioxide snow that accumulate in grooves and at the bases of steep slopes in colder areas on the mountain. At upper right is a global view of Io showing the location of the mosaic (Image courtesy NASA/JPL/University of Arizona/Arizona State University; additional processing by the author)

assure us that by the time humans make landfall on the violent moon, our maps will look very little like they do today.

As for a safe harbor on the hellish little world, John Spencer says the leading hemisphere would be a bit more sheltered from radiation, as Jupiter's plasma comes up from behind, "but I don't think anywhere would be safe except underground, and then you'd have lava to worry about. It's great to look at from a distance; it would not be fun to be immersed in it. Still, all these dangers are very interesting. The view from Io would be incredible."

When he's not staring down the throats of active volcanoes, Ashley Davies is a Research Scientist at the Jet Propulsion Laboratory, Pasadena, California. Here's what he had to say about the possibility of visiting Io:

Fig. 5.7 Io (JPL)

The thing that brought me from the UK to JPL – to the United States – was to work with Dennis Matson and Torrence [Johnson] on analyzing their ground-based observations, to really understand the volcanic processes taking place. This was in 1994, when we were using the Infra Red Telescope Facility on Mauna Kea (Hawaii). Galileo would not arrive [at Jupiter] until the next year.

What intrigues me [about Io] is the incredible level of dynamic activity that is taking place, the incredible range of volcanic activity, the largest, most powerful eruptions that we've ever seen anywhere on a planetary body. The eruptions cover eight orders of magnitude. One thing that isn't understood is what happens at the small end of the scale. Just about every volcanic eruption that's seen on Io is larger than the largest eruptions seen on Earth. We don't know if it's a function of resolution or instrument sensitivity, or whether it's a function of Io's volcanism and limits imposed upon the transport of magma from interior to exterior. We're trying to investigate this with current [Galileo] data. We're just not seeing small eruptions. One thing that is an advantage over remote sensing is being up close and personal to see the intricacies taking place. Of course, the technological hurdles that will have to be overcome before it's worth sending somebody to such a hostile environment are considerable. It is such a dynamic system, and the geological processes are taking place on such astounding scales.

If technology was no limit, I'd want to go stand on the shores of Loki and send some grad student down to investigate close up. You need to have a good supply of grad students when you go investigate Io. I would like to examine the time-series of the events: how eruptions start, how they build up, how they die down, how they interact with the volatiles in the crust and on the surface. What exactly is on the surface? What is the structure of the surface? I'd like to investigate the walls of paterae, the mountains and the thrust-faulting that forms them. Mostly, I'd like to investigate how volcanoes work.

I do want to go to Loki Patera to determine what the primary resurfacing mechanism is. Imagine walking up a shallow rise towards the edge of the patera, reaching the edge and seeing, stretching towards the horizon… seeing … um… yikes. Wow! So that's what's going on! Cue dance, celebrations, champagne, publications.

EUROPA

Europa is the poster child for extraterrestrial oceans, and the search for extraterrestrial life.

Jupiter's fourth-largest moon sparkles with a brilliant white surface made of glistening water-ice, inscribed by fractures and faults. Its pristine surface points to a very young age, only about 50 million years by some estimates. An elegant calligraphy etches the surface in linear and bow-shaped stripes – telltale signs of powerful tectonic forces. Researchers have constructed several models to describe conditions beneath Europa's bizarre, grooved facade, ranging from soft ice to an ocean 100 km deep. Whatever its true internal form, Europa is far more a "water world" than Earth ever was.

High-resolution images taken by the Galileo orbiter offer compelling circumstantial evidence of an ocean on Europa. Ruler-straight lines streak across the frozen landscape, bracketed by long ridges rising hundreds of feet into the black sky. The ridged surfaces have fractured into vast sections of ice rafts. These rafts have apparently shifted and rotated before freezing solid again. Still other areas, called chaos regions, seem to have collapsed into a sea-like slurry, freezing into place after fracturing into puzzle pieces, just waiting for scientists to reassemble them.

Aside from visual clues, Europa generates a magnetic field consistent with liquid salt water. In the first week of 2000, the Galileo spacecraft flew within 346 km of Europa's surface. As the craft braved deadly radiation fields from nearby Jupiter, its magnetometer sensed a change in magnetic fields coming from Europa. To experts in fields and particles, these directional changes looked hauntingly familiar – they resembled those generated by Earth's oceans, just the kind that would be generated by electrically conducting liquid within Europa's upper ice region.

Unlike the electrical currents pouring from Earth's core, Europa's field is induced – it is created in response to Jupiter's prodigious field lines. This induced field constantly changes in response to the rapidly rotating magnetic field of Jupiter. While Europa circles the giant planet in just over 3.5 days, Jupiter spins on its axis once every 10 h. Its magnetic field is tilted from its spin axis by nearly 10°, and Jupiter's pace swings its field around with it, sweeping through Europa. Under these brutal conditions, any conducting material swept by this field will create (or induce) a magnetic field to offset the rapidly changing external magnetic field. This induced field is exactly what scientists observed with the Galileo spacecraft as it flew past Europa. The presence of the induced magnetic field led researchers to the conclusion that a near-surface conducting layer, such as an ocean with dissolved salts, was the culprit.

Europa travels halfway around Jupiter each time that Io completes an orbit, and twice for each of Ganymede's circuits. This means that in terms of the tug of gravity from its siblings, Europa is caught between Io's "rock" and Ganymede's "hard place." Tidal heating – generated as a result of the moon's gravitational tugging – heats up Europa's interior, though to a much lesser extent than Io's. Conventional volcanism may exist on Europa's "sea floor," where the silicate mantle meets the ocean above. Europa may have a volcano-rich sea floor simmering at the bottom of an oceanic abyss.

Although submerged in eternal darkness, this stygian environment has one advantage for any Europan life-forms: it is sheltered from the fierce radiation of Jupiter. The ice crust provides a barrier to the deadly rain of energy bombarding Europa. But the surface lies bare to this onslaught.

The Jovian magnetosphere acts like a giant particle accelerator. Jupiter's magnetosphere drapes the surface in enough radiation to tear apart any cell walls within moments. Even an unshielded astronaut would receive a fatal dose of radiation in a day.[1] But some areas on Europa are more sheltered than others. Because Jupiter rotates more quickly than Europa orbits it, the planet's magnetosphere actually overtakes the moon (just as it does with Io). Its super-energized particles constantly sweep by the little moon, bombarding it from behind, so the most extensive radiation falls in the trailing hemisphere area. The safest places to go are high latitudes on the leading hemisphere or the sub- and anti-Jovian high latitudes (see Chapter Ten).

LIVING ON THE EDGE

One of the most compelling arguments for Europa exploration is the possibility of life there. It was only in 1977 that explorers discovered sea floor volcanism on the ocean floor along the Galapagos Rift zone near the Galapagos archipelago. Scientists had noted hot undersea plumes, but their nature and source remained a mystery until the deep-sea submersible *Alvin* revealed dramatic chimneys of sulfur compounds rising from the ocean floor.

In the years that followed, researchers came to realize that the number of volcanoes on the ocean floor must dwarf the 500–600 active ones on the surface. Undersea hydrothermal vents break through along the mid-ocean ridges where new crust is being created in Earth's stony conveyor-belt of sea-floor spreading. Although undersea volcanism is not limited to plate boundaries, mid-ocean ridge areas may well be the most volcanically active sites on Earth.

Vents tend to cluster in groups, much as they do in Yellowstone Park in the United States. Earth's mantle comes to within a few hundred meters of the surface in these eternally dark regions. Seawater percolates through Earth's crust, eventually making contact with the 1,200 ° C magma. The water is able to heat up to 540 °C, because the high pressure prevents it from boiling. The heated fluid makes its way up through fissures in the rock, leaching minerals along the way. When it finally flows into the ocean, it is laced with a complex mineral soup. Mineral-laden water streams from these sources, building delicate structures of spires and chimneys, some of which may tower dozens of feet above the sea floor. The streams of water are often charged with materials that lend names to their appearance. Some vents are known as black smokers, while others are called white smokers.

Not all of the undersea extremophiles are hot-water inhabitants. Europa's sea floor may well have active volcanism, too. The biomes at sites such as the Galapagos Rift zone are completely independent of a Sun-centered food chain. Their energy comes from the witch's brew of chemistry emanating from the volcanic vents. Perhaps Europa has developed a similar biological chain, working its way from searing water outward. In the depths of a 100-km ocean, the eternally dark environment sheltered from Jupiter's deadly radiation, the complete lack of sunlight might be offset by the rich minerals of the sea floor vents. If so, future diving vessels will undoubtedly be dispatched to view the eruptive sites, where we may find the first alien life in our Solar System.

What of surface activity? At the time scientists were sifting through Galileo data, they found many features suggestive of ice volcanoes or geyser-like eruptions. Dozens of sites on the moon hint at past eruptive events involving water that has rapidly frozen in Europa's near-vacuum environment. One of the clues leading to this conclusion lies within the "triple bands," highway-like parallel stripes that contribute to Europa's cracked eggshell appearance. Called linea, these uniquely Europan formations consist of bright lines running down the center of a dark, well-defined band. The bands are less than 15 km across, but run over Europa's face for thousands of kilometers.

Fig. 5.8 Left: Multiple sets of ridges cut across Europa's flatlands. Several faults, or fractures, have moved parts of the landscape horizontally. Right: On its closest flyby of Europa, the Galileo spacecraft snapped this image of the moon's rugged, bizarre landscape. The image covers an area roughly 1.8 km across. Dark material has gathered in the hollows of the bright ice crests. Across the center of the image is a disarrayed assortment of hills nested between Europa's famous ridged terrain, top and bottom (Images courtesy of NASA/JPL)

The linea are directional fractures. In the northern hemisphere, they trend in a northwest direction, while southern fractures trend southwesterly. This directional tendency suggests a relationship with Europa's orbital stresses. Galileo's imaging system – superior to those of the Voyagers – revealed the borders of the bands to be diffuse and irregular in many areas. Even the central bright median displayed patchy sites with halos of bright material spilling across the dark outer band. The triple bands gave the appearance of fissures erupting materials of varying albedos, or levels of brightness, onto the moon's sparkling surface like a beaded necklace. Were these remnants of escaping water vapor? The fingerprint of past geysers?

One leading theory for triple-band formation proposes a tidally induced fault breaking through the ice to the ocean below. A "cryolava" of briny water oozes up to seal the vent, while geyser-like plumes erupt from weaker locations. This style of cryovolcanism is referred to as "stress-controlled cryovolcanic eruption." The water eruptions coming out of linear fissures may be analogous to rift-magma eruptions on Hawaii. As the region around the fracture builds vertically, the weight of the growing ridge pulls on the surrounding ice, causing parallel fractures. These cracks, in turn, develop into more parallel ridges, duplicating the process as the band expands in girth.

Other planetologists suggest that linea were emplaced by solid ice rather than liquid breaking through the surface. They cite several reasons, including the fact that many of the bands rise hundreds of meters above the surrounding plains. This suggests ductile ice rather than liquid (they reason that liquid would not result in a raised structure). The genesis of domes and hummocks, common in some areas, also makes more sense in this scenario, they reason. Furthermore, there seems to be no flooding of material into adjacent ridges and valleys, which would have occurred had the linea been filled with liquid.

Ridges usually form as a result of compression, the pressing together of two segments of the crust. Fractures form as a result of stresses that crack the brittle crust, pulling it apart. Both tectonic movements are common on Europa.

How thick is the crust? Some studies suggest an extensive, deep ice shell around Europa, essentially solid down to many tens of kilometers. Other models posit an ice crust of something like 10–15 km. The structure and thickness of Europan crust is a hotly debated issue, and a complicated one at that. Some researchers assert that impact craters and jumbled "chaotic zones" indicate a thin, 2-km crust at the equator. The crust appears to thicken to the north and south. Much of the evidence indicates that Europa has a global ocean, a vast subsurface sea running from pole to pole under the ice. Scientists are able to piece together a picture of Europa's interior by the way Europa's gravity affected the path of the Galileo space-craft. As the spacecraft sped up, its signal shifted, just as the siren on a passing fire truck shifts. The varying structure of Europa's crust caused subtle changes in this "Doppler" shift, enabling scientists to chart the moon's internal structure. Doppler data from the closest flybys fits a rocky interior capped by an outer layer of water 100–200 km deep. A few-kilometer-thick crust is a fragile film over such a massive ocean.

Some research suggests that Europa's subsurface oceans may remain liquid because of another kind of tidal friction. Because of its small obliquity (non-circular orbit), the gravity of Jupiter may set up tidal waves called Rossby waves. These slow-moving undulations

Fig. 5.9 Telltale signs? Some researchers believe the features in Astypalaea Linea point to folding of the surface through compression. Ice appears to have spread and cracked in one area, while to the north ridges have risen up (Galileo images courtesy of NASA/JPL, additional enhancement by the author)

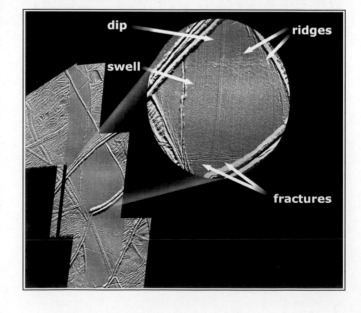

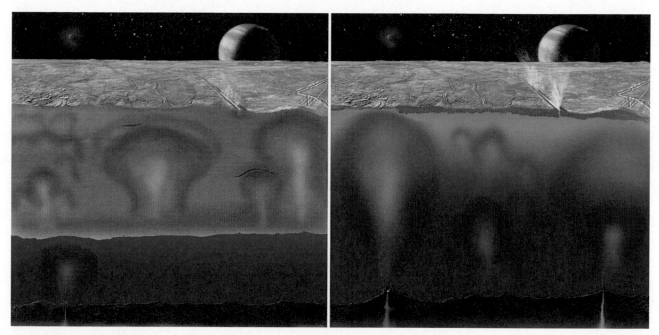

Fig. 5.10 *Researchers attempt to explain Europa's surface features by either a thick ice crust overlaying a shallow ocean (left) or a thin one with a deep ocean (Image courtesy of NASA/JPL; art by Michael Carroll)*

can generate significant energy. In fact, estimates range as high as 2,000 times the energy that Europa receives from tidal heating alone.

Conjectures and theories about escaping water vapor at Europa came to the fore in late 2013, when a series of Hubble Space Telescope images revealed the signature of water hovering over the southern hemisphere of Europa. Given the amount of water in the cosmic cloud around Europa, some investigators estimate the moon is erupting a stunning 7 tons of water every second. The great water plume reaches some 200 km high, and may be the fingerprint of geysers erupting at an estimated 700 m per second, or three times the speed of a passenger airliner.

Scientists are awaiting further confirmation, but the purported discovery would demonstrate that liquid water is close to the surface. It may lie in localized ponds, or the vapor/geysers may be direct links to the deep ocean below. Either way, if the eruptive sites can be located, they will be prime targets for tomorrow's astrobiologists seeking answers to what lies in those oceans beneath Europa's icy landscape.

Although water-ice is the dominant substance on the surface, instruments have detected other important materials mixed in with the ice, and some of the data suggests that these materials may come up from the ocean below. Amorphous, often radial discolorations may be the fallout from geyser-like activity like that

Fig. 5.11 *Blue pixels overlaying a Voyager image of Europa show the HST's detection of water vapor floating over the southern polar regions (Image courtesy of the Space Telescope Science Institute)*

possibly detected by Hubble. Such stained sites are called "painted terrain." Many such features exist on Europa. Most are associated with fractures or faults. The brown to rust stains appear to be endogenic (generated from the inside), but whether they are localized events or global in nature remains to be determined. Some have well defined, flow-like or pooled edges. Others are diffuse, fading at the boundary. Often dark material blankets the surrounding terrain, more like particulate matter than a flood of liquid.

Just what is this dark material? Instruments aboard the Galileo spacecraft gave researchers enough data to identify several candidate substances. Galileo's infrared (heat) data suggests various salts, with magnesium sulfate (like Epsom salts) or chloride salts as the best spectral match. However, some investigators argue that Europa's spectra is better matched by hydrated sulfuric acid. Although salts are not brown, sulfur is. The inner Jovian environment is bathed in sulfur, thanks to Io's eruptions. Clays have also been teased out of the data. These minerals were likely delivered by asteroids or comets impacting the surface. Other scientists have suggested iron compounds that presumably issue from the rocky core.

Louise Proctor is a planetary scientist at Johns Hopkins University Applied Physics Laboratory in Laurel, Md. She says "Europa does have material moving up thru the ice shell like diapiric upwelling, sort of cryo-volcanic material coming up and punching through the surface. Some of that has frozen above the surface, hasn't yet relaxed or been pounded down by impacts. You could do bobsled or the luge. A slow-motion luge in that low gravity, and you'd be out of the radiation since you'd be in between ridges. There could be a whole winter sports angle to this we haven't explored. How about the 2064 Olympics? Maybe I'm wasted in my job; I should go into marketing or something."

The reddish material discussed above is darker than the surrounding landscape and may be the expression of briny subsurface lakes. Radiation also tints the surface ice in rusty tan or greenish-gray, especially in the trailing hemisphere. One researcher quipped that Europa "has an unhealthy tinge." Once the surface is stained, it brightens over time. Older linea are nearly the albedo of the surrounding plains on which they lay.

Given Europa's orbital stresses, many planetary geologists believe it likely that Europa's ocean floor is peppered by volcanic vents. Their great depth might preclude any direct disruption of the surface, so Europa may not exhibit any visible clues about sea-floor volcanism. Still, there are mysterious features hinting at forces beneath the crust. The most striking of these are the chaotic regions. Here, ridges are seen to slip across each other in lateral faults, while remnants of ridged ground have broken into rafts and rotated in a slush of debris. Many of these rafts fit together like

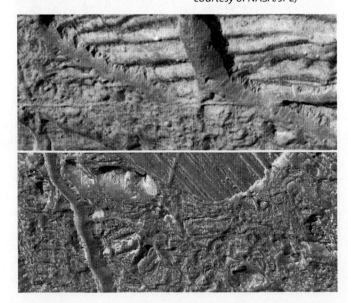

Fig. 5.12 Two views of chaos. These detailed views give an appreciation of the rugged, jumbled landscape of chaos regions on Europa. The top view shows two ridged units of crust that have pulled apart, while the bottom image shows a disintegrated landscape abutting ridged plains at the top (Images courtesy of NASA/JPL)

pieces of a cosmic jigsaw puzzle, clearly indicating that a once continuous surface has been split up and moved around. Chaotic terrain often coexists with relatively smooth, craterless landscapes. Chaotic terrains bear a striking similarity to sea ice in Arctic regions. In the terrestrial case, solid ice has fractured, drifted into new positions, and been frozen in place again.

Some researchers suggest that plumes of hot water, generated by sea-floor volcanic sources, could thin the ice, eventually triggering a series of melt-through events. The ice would fracture, freeing rafts of surface ice to bob and rotate in the quickly solidifying lake. If the concept is correct, the rafts should rotate in the same direction as the hypothesized plume (clockwise in the north). This seems to be the case, although data is limited.

It is also possible that chaotic regions have been generated more indirectly. What concerns many planetary geologists is the distance between Europa's ocean floor and the base of its surface crust. If models are correct, plume material would take weeks or even months to migrate from the sea floor to the surface. Additionally, chaotic regions would need to be heated for extended periods of time to explain the crustal movements observed.

Some researchers have solved the distant plume problem by proposing warm columns of ice. In this scenario, a heated plume warms the ice over a long period. The heat moves through the ice much as the dayglow material in a lava lamp. Rather than melting completely through the ice, the process would be gradual and relentless, softening the ice enough to free the rafts for extended periods of time. Additionally, impurities in the ice might help the process along by the lowering of the melting point. Small amounts of salt or sulfuric acid – both of which have been tentatively identified by Galileo – could provide enough force to generate the domes, even through tens of kilometers of ice.

A widely accepted scenario theorizes that while the ice is heated from beneath – possibly from sea-floor plumes – no melt-through occurs. Instead, a rising solid mass (called a diapir) makes its way through the crust, eventually reaching the surface. These diapirs could also interact with pockets of trapped salty water. Chaotic regions need not be generated quickly, but could be the result of a long and gradual process.

The diapir model would help to explain the lenticulae – dark spots or areas. Lenticulae come in varied forms, including miniature chaos areas, pits, depressions and domes. The term is Latin for "freckles." Many of these dark spots tend to be depressed beneath the surrounding plain and are stained. The ruddy ice is briny, bolstering the idea that dark materials are seeping – or erupting –onto the surface.

The same forces that robbed Io of its water and turned its interior molten have had a

Fig. 5.13 Top: Twin areas of reddened terrain, Thera and Thrace Maculae, erupt onto the older icy ridged plains of Europa. Thera (left) is about 70 km wide by 85 km high and appears to lie slightly below the level of the surrounding plains. Within it, icy plates have broken free from the edges of the surrounding chaos region. Thrace (right) is longer and appears to dome above the older surrounding bright plains. Bottom: Agenor Linea spreads a bright band across Europa's icy face. Imaging scientists have draped high resolution color images of Agenor over lower resolution regional imagery. Agenor is a "triple band," flanked by dark, reddish material (Images courtesy of NASA/JPL)

drastically different effect on Europa. This ice world, with its curving fractures, parallel ridges, and jumbled landscapes, offers a complex world for future travelers to explore. Given the right equipment, they may even be able to go for a swim.

Robert Pappalardo is a Senior Research Scientist in the Planetary Science Section, Science Division at the Jet Propulsion Laboratory in Pasadena, California. While Ganymede is his favorite of the Galilean satellites, Europa is a close second. He says this about why he'd love to see Europa up close:

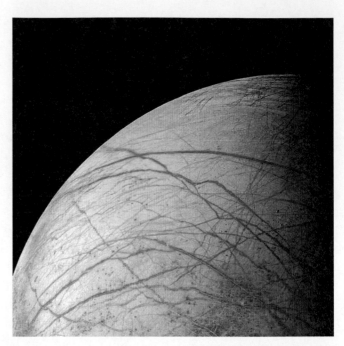

Fig. 5.14 Bands across Europa (JPL)

The Voyager Jupiter missions of the late seventies initially sparked my interest in the Galilean satellites. I still have the Astronomy magazine article "Four New Worlds" by Torrence (Johnson) and others. When I was a grad student, I worked with Ron Greeley. He was assigned the action of planning the Europa encounter imaging by Galileo. I remember him coming back from a meeting and saying "They divvied up the satellites and we got the jewel in the crown, Europa." I had just finished my Ph. D and worked with Ron on some of that planning. In doing so, we were looking at the Voyager images, and that led to finding evidence of the pull-apart geology of Rhadamanthus Linea. I love Ganymede as well, but Europa, for me, is a little more mysterious. Ganymede, we kind of get it. There aren't these emotional arguments at meetings, or these big mysteries. There are wonderful things about Ganymede, but Europa we still don't really understand.

It's hard to beat that view of Jupiter up in the sky, changing and rotating and morphing. I was imagining putting my little lawn chair out there on Europa somewhere. A low radiation zone would be nice, somewhere on the Jupiter-facing part of the leading hemisphere, off north or south a little bit. I'd say north, so things aren't upside down. I'd want some nice chaos terrain; a big cliff on the right somewhere, and on the left maybe a big ridge stretching off to the horizon. And maybe a nice plume puffing away somewhere nearby.

I live near Venice beach. It's exactly 2 km away. That's the width of a typical double ridge on Europa. They're about 100 m high and about 2 km wide. That's a half hour walk, and 100 m is not very high, but a 10° slope is fairly respectable. In some places it reaches around 30°, close to the angle of repose. It would still be quite a hike up that thing and down into the central trough, where there might be some erupting going on. Just north of us is a beautiful valley called the Carrizo Plain. The San Andreas Fault cuts through it. The San Andreas has very distinct topography there. It rises up; there's a valley where the fault itself is along that rise, and it drops down again. It's a double ridge.

That's how I imagine Europa. On Europa it would be higher still. The dip in the middle is a couple of football fields wide. I think it would be rubbly. We see evidence of blocks at the bases of the ridges, where the stuff has fallen down, and when you see slopes around 30°, it suggests angle of repose, [which would mean that] this unconsolidated debris is forming that slope. It might be a bit bouldery, especially up near the top. You'd need some good shoes.

Europa would have a spectacular view of Jupiter, and it's the place where a lot of the astrobiologists want to go. If you're looking for life, Europa's the place.

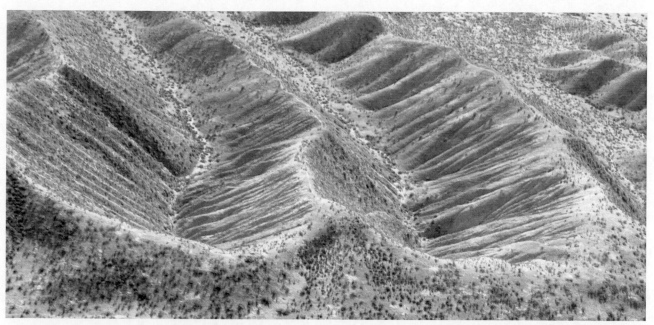

Fig. 5.15 *The San Andreas Fault provides a close analog to the double ridges found on Europa (USGS photo by David K. Lynch, Kenneth W. Hudnut and David S. P. Dearborn, 2009. Low altitude aerial color digital photographic survey of the San Andreas Fault in the Carrizo Plain)*

GANYMEDE

The big kahuna for the Jovian moons is Ganymede. Measuring 5,268 km across, its diameter bests that of the planet Mercury by nearly 400 km. Ganymede is a planet in its own right. It is the only moon known to generate an internal magnetic field, implying a hot convecting core containing molten iron. Spacecraft have also found evidence of an induced magnetic field like Europa's, suggesting a deep internal ocean of liquid briny water. The densities of both Ganymede and Callisto suggest that the two large satellites contain about 60 % rock and 40 % water. How much of that water is ice, and how much – if any – constitutes a deep liquid ocean, is open for debate. Because of the massive amounts of ice in their makeup, the moons are less dense than Earth's own, creating gravity fields of about 6/7 that of Earth's Moon, or 1/7 Earth gravity.

Like that of its siblings Europa and Io, Ganymede's geology is complex, displaying a marked contrast in its terrains. Dark terrain, heavily cratered, forms about one-third of the surface, while the other two-thirds consist of swaths of bright, grooved terrain that may have been formed by both cryovolcanic eruptions and tectonic forces. In the aftermath of the Voyager Jupiter encounters, a consensus arose that Ganymede was frozen in its evolution. Proctor, the senior planetary scientist at the Johns Hopkins University's Applied Physics Laboratory, says:

> I like to think of Ganymede as almost starting to turn into Europa. The old, heavily cratered surface began to rip apart or pull apart in some way, but then it stalled. It's kind of in limbo, halfway between Callisto and Europa. It hit the sweet spot

between Europa, where everything has been happening and there's been a lot of upheaval and may still be going on today, and Callisto, that's been quietly sitting there as this witness place for the Solar System almost since its inception. At Ganymede, you can study a really old terrain that's relatively unchanged, but it's also a wonderful place for structural geologists to go and play because it has this incredible surface: two thirds of it has been ripped apart by these fantastic, gigantic lanes of icy grooves, and we still don't completely understand how they formed. What caused Ganymede to undergo this massive upheaval?

The dark, ancient terrains resemble those of Callisto, while the bright regions are somewhat like Europa. But with new data from the Galileo mission, ice moon gurus realize that Ganymede is more complex, its history not so simple.

Ganymede's brown-gray landscapes appear to be ancient, battered surfaces covered in loose dark material with a few icy slopes scattered throughout. The dark terrain reminds Geoffrey Collins of the dark terrain on Callisto, the moon next door. Collins should know. He's been studying Ganymede's dark terrain for most of his career. "The dark terrain seems to be covered in this dark dust. You have bright knobs of ice sticking out of the dark regolith."

Ganymede's dark terrain is heavily cratered and thought to be ancient, older than 4 billion years. Its composition shows a greater fraction of rocky material than the icier bright terrain. The dark terrain has been modified by numerous geologic processes, including tectonics. Geoff Collins has been studying the most detailed images available, taken by the Galileo spacecraft. "There were a handful of images in the 20-m or better range. A couple of them are in dark terrain, and you see this dark dust with these bright donuts of ice where craters rims are sticking out." These images seem to show that the dark terrain is thin, perhaps a blanket of material layered over fresh ice.

The bright terrain transitions from dark ancient regions abruptly, Collins says. "Often when you see the transition between one and the other, it's like this old, heavily cratered surface just stops, and there's this cliff. You look out and there's this field of line upon line of icy ridges and cliffs." The bright terrain tends to be lower than the dark terrain. It is probably at least 2 billion years old, so craters pepper its surface, but its crater count is far less than that in the dark terrain.

Parallel valleys and ridges, called grooved terrain, plow across Ganymede's bright regions. Tens of kilometers wide, these folded track-ways of ice run across the face of Ganymede for hundreds of kilometers, slicing swaths through the dark terrain, breaking it into gigantic polygons. The ridges themselves are not made like many mountain chains on Earth, where plates shove together and force wrinkled ground upward. Rather, these valleys look more like slabs of ice that have broken apart along fault lines and leaned over. "Each ridge is not being forced up like the Himalayas," says APL's Proctor. "It's being tilted like books on a bookshelf. Radiogenic heat may have driven those grooves."

It appears that the forces of gravity from Jupiter and the other Galileans have stretched the massive moon's ice crust. This extensional

Fig. 5.16 *This view is one of the highest resolution photos of the boundary between dark and light terrain. Dark terrain to the left transitions to lighter terrain at right. Bright ice rubble fills canyon floors where faults cut through, while remnants of the ancient dark landscape are preserved on the ridge tops. Some craters are old and softened; in other places donuts of bright icy rings protrude from a dark mantle. Landslides have left rivulets and streamers on the slopes. To the right, the material starts to change character. The old, softened craters are gone, and there are fewer craters overall (Image courtesy of NASA/JPL)*

8. The word *graben* comes from the German "trench"; the related "das Grab" means "grave." In the Middle Ages, many European cultures buried several people on top of each other to conserve space in cemeteries. Commonly used pine coffins would collapse under the weight of passersby, causing the ground to sink, like a geological graben.

9. Sublimation happens in a vacuum or low pressure environment, where water or gas melts directly from ice into vapor.

faulting could also cause graben,[8] valleys resulting from the downward slip of the surface. Several high-resolution Galileo images detail the bright ridges wandering across the surface, but the structures have been modified over time, eroded by small craters and softened by avalanches. Piles of boulders and other debris line the base of the cliffs. Some of the ridges look as though they are in the process of disintegrating into linear piles of rubble. Whatever made the bright terrain did it a long time ago.

Although some of the ridges on Ganymede resemble those of Europa, their cause is quite different. On Europa, long belts of ground have been pulled violently apart. As they spread, new material wells up to fill the gap. We can see evidence of this process in places where one pattern of ground is continued on the other side of one of these bands. But in an apparent attempt to avoid Europan redundancy, Ganymede has no such repetition of form (except in a few rare cases). Instead, the dark terrain of Ganymede appears to have stretched as it broke apart, and was then somehow transformed into bright terrain, Prockter observes. "Light terrain is grooved. At lower resolutions it looks smooth, but we've since discovered from other bodies – and better imaging – that there's very little that is really smooth. Bright terrain has these big parallel lanes of material; they may pinch and swell but mostly they're very linear. We see the grooves going down almost to the limit of resolution. You keep zooming in and zooming in and you just keep seeing more of the same. A lot of the lanes are large, regional, going on for hundreds of kilometers." The composition of Ganymede's bright terrain is almost pure water ice, suggesting cryovolcanism or flooding from the interior, but its ridges and valleys imply that tectonic forces are to blame. Some evidence hints that the dark material is not dark all the way through. Rather, the regions are simply dirty ice material where sublimation[9] from the surface creates a dark lag (material left behind). Under this scenario, tectonics then come along and slice that surface up with extension, exposing cleaner ice.

This, in fact, is one of the mysteries yet to be solved on Ganymede. It faces researchers in the form of a choice: Does the bright terrain form by ripping apart the dark terrain, exposing the bright ice that's underneath, or do new liquid water "lava flows" flood the surface, covering the dark terrain? Investigators like Collins and Proctor are baffled. Proctor says, "With the Galileo mission, we expected to solve the riddle of how the grooved terrain forms, and although we know extension plays a very major part in its formation, there's still some ambiguity about how the material gets to the surface – if it's new material or whether it is simply dirty, dark terrain that has been fractured so that new clean material is now open and exposed." Geoff Collins adds, "There are places where it could be argued either way, and maybe there's not only one answer to that question. That's something that I still find intriguing, and something I hope someday to have enough data to resolve. Someday."

A visual survey of Ganymede reveals both similarities and differences with Callisto and Europa. The illustrations here make those clearer.

Like all planets and moons, Ganymede has been battered by asteroids, comets and meteors. Craters left behind offer clues to what the surface is made of and what may lie underneath. Impacts have scarred Ganymede's face on many scales, leaving features ranging from small bowls a few feet across to the vast impact basin of Gilgamesh, some 800 km in diameter. Small craters sometimes dig out dark floors or cast rays of dark material. This darkening might be caused by solar radiation darkening impurities in the ice, or it may be material left over from the impactor. Larger craters seem to burst across the moon's dark blue-gray surface with splashes of bright ice. As craters age, these rays fade.

The centers of many mid-sized and large craters hold strange central pits. These structures may be the result of warmer ice welling up from inside after the impact, or the backsplash of the impact itself. The larger craters have a mount inside the central pit, adding to the mystery.

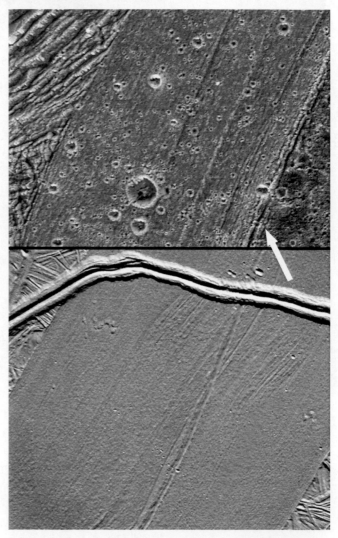

Fig. 5.17 The Arbela Sulcus region on Ganymede (top) resembles areas on Europa like Thynia Linea (bottom), where the crust has separated and been filled in with new material. These images are the same scale. Note the Europa like ridge system (arrowed) on Ganymede, similar in structure but not in scale. The youth of Europa's surface is evident by its lack of craters (Image courtesy of NASA/JPL/Brown University)

Fig. 5.18 Ganymede's Galileo Regio (top) and Callisto's vast Valhalla impact basin both show concentric wave forms. Note that the irregular horizontal central strip on the Ganymede image is due to missing data (Images courtesy of JPL)

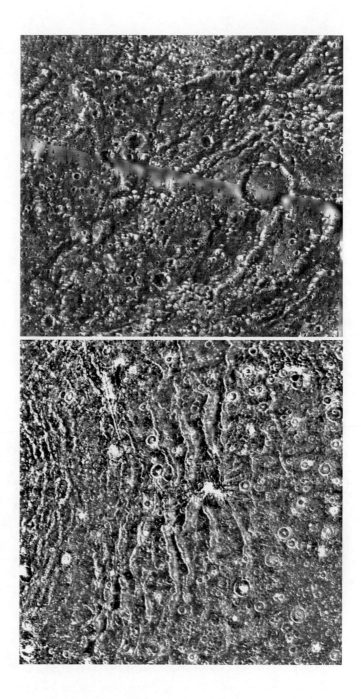

As Ganymede's craters erode, they break into icy knobs rising from the dark regolith. Frost sometimes accumulates against shadowed slopes. Ancient crater rims and eroded ridges take on the appearance of foam floating upon a dark sea.

Across many of Ganymede's frozen plains, the ghosts of craters haunt the wilderness. Scientists call them palimpsests. The name hearkens back to a time when writers penned messages on parchment or vellum. These materials were scarce and expensive, so when the piece of parchment had served its purpose, writing could be scraped from the surface, often using

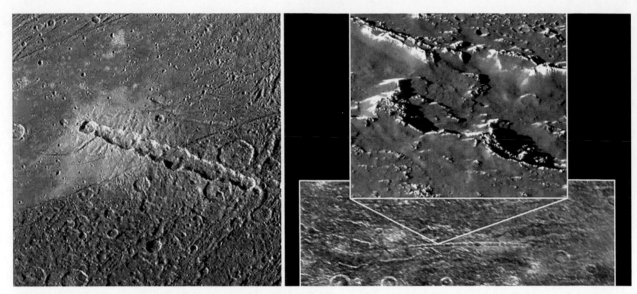

Fig. 5.19 In December of 1994, Comet Shoemaker-Levy 9 passed close to Jupiter and fragmented into over a dozen pieces. Months later, it slammed into Jupiter, peppering the Jovian clouds with dramatic, Earth-sized donuts of smoke from the explosions. Evidence of other such events exists at both Ganymede (left) and Callisto in the form of crater chains (Images courtesy of JPL)

Fig. 5.20 Ganymede's crater erosion seems to mimic Callisto's in a subtle form. In both cases, ice rims and edges of high terrain stand above darker plains material (Images from the Galileo mission, courtesy JPL)

milk and oat bran. The parchment would then be used again, but over time, the old writing bleeds through.

Just like an old parchment, Ganymede's ancient narrative comes subtly to the surface. Palimpsest craters likely formed early in Ganymede's formative years, when meteors broke through the rigid ice crust, blasting out slushy material over grooves that can still just be seen near the edges of the ghost craters, valleys of the past frozen beneath the palimpsest's surface. Within the palimpsest, craters have punched through to darker material.

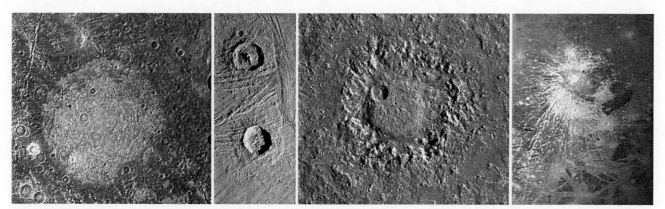

Fig. 5.21 Far left: The palimpsest Memphis Facula spreads a light pond of ice 350 km across the ancient landscape of Ganymede. Note the grooved terrain showing through near the edges. Center left: Small craters on Ganymede tend to have central peaks, and often rest upon a raised "pedestal." Center right: Large craters develop pits and domes at center. Similar craters also form on Callisto. Far right: Rayed craters scatter fresh ice across older, darker surfaces (All images courtesy of JPL)

The form of these craters suggests that they are thin disks of lighter material.

Many craters have been split, pulled apart, and otherwise obliterated by the shifting of great "plates" of surface ice. But the planet-sized moon arrived at its fractured countenance via a far different route than Earth did. Earth's plates subduct (go under each other), expand in sea-floor spreading, or compress and fold to form mountain ranges. While one part of the planet extends, another part is contracting, evening out the globe's surface area. On Ganymede, the bright grooved terrain seems to be extensional only, meaning that the grooves are scars of planetary expansion, rather than simply local expansion. In short, Ganymede swelled as it froze, cracking its surface.

Ganymede's daytime temperatures are far more extreme than anything in our Earthly experience, averaging −140° C. But at the poles, the moon's temperatures plummet even farther. Here, a dusting of frost drapes across craters and valleys, covering them in a thin veil. Researchers suspect this frost is primarily water-ice but may be mixed with carbon dioxide ice as well, says icy satellites expert Jeff Moore. "Ganymede, with all of its local hills, especially in the old, cratered dark terrain, will have some deposits of ice on it which might be a few meters thick. So it might look like it snowed up there on those hills. Here we see fairly substantial deposits of frost and ice in the polar regions. [With the Galileo spacecraft] we did take at least one strip that ran along at high northern latitudes, and we saw that the cold-facing slopes had substantial, discreet ice deposits in them. That's what gives Ganymede those bright polar hoods."

Aside from temperature, another alien factor contributes to the poles. The polar ice builds up in regions that line up with the gaps in Ganymede's magnetic field lines. In these polar areas, Ganymede's magnetic field does not protect its surface from the inflow of Jupiter's harsh radiation. In effect, Ganymede's magnetic field has a direct expression on its surface geology.

Like Europa, Ganymede has an ocean lurking within. In the case of Europa, the ocean rests upon a floor of rock, where minerals and perhaps even volcanoes add life-enabling materials to the mix. Not so at Ganymede. The moon's induced magnetic field portends an interior ocean beneath the frozen crust, but it is probably sandwiched between ice layers above and below, sloshing some 170 km beneath the surface. Cut off from the mineral-rich rocky core, what chance does this ocean have of sustaining life? Several recent studies offer results that give some astrobiologists pause. Models indicate that salty water at the ice/rock interface may percolate upward, migrating toward the ice-locked ocean above. The water may congregate in ponds along the way. The pressures and temperatures found in Ganymede's ice are difficult to simulate in the lab, but it may be that the ice barrier between ocean and rock is not such a barrier after all. And while they have found no shopping malls, astrobiologists are beginning to reconsider Ganymede as a possible site of life.

Ganymede also has something unique among the icy satellites – an intrinsic magnetic field generated by a molten core. "This is the most important thing about Ganymede," says Louise Proctor. "It has its own magnetosphere, and it's really got no right to have its own magnetosphere that it's generated itself. The only other two places in the Solar System that have their own magnetospheres (aside from the giants) are Earth and Mercury. So Ganymede, weirdly – of all the moons – is still active in its interior sufficient that it can stand off the Jovian magnetic field. That makes it a very unusual body." For human exploration, this distinction is important. Ganymede's magnetic bubble provides at least some protection from the fierce radiation of nearby Jupiter, cutting down on the need for habitat shielding, radiation-hardened equipment, and bulky suits.

Geoff Collins imagines what travelers in the future will come upon as they hike the long crags of Ganymede. "The dark terrain might be the equivalent of the lunar highlands, except with these amazing albedo contrasts of the bright, icy slopes and the dark, dusty material down in the bottoms of craters and troughs." He continues: "The bright terrain is newer, but it's still over a billion years old, so it's had the edges worn off. I think about those Apollo pictures – those soft-looking mountains – and some of the things on Ganymede might look like that up close. On my wall I have a picture of Hadley Rille when *Apollo 15* was up there. From a distance, Hadley Rille looks pretty sharp; there are these nice flat plains and then it abruptly dips down into this valley. But when the astronauts got up close, there was maybe a little rock exposed along the edges, but a lot of it was covered in rubble and debris and dust. That's how I imagine Ganymede looking."

Geoff Collins has been involved with various NASA projects such as the Galileo mission to Jupiter and the Cassini mission to

Fig. 5.22 Galileo Regio, Ganymede (JPL)

Saturn. Here is what he had to say about what he would like to see up close and personal in the outer Solar System:

> I come from a geology background. I fell in love with geology first, and I was always interested in faults and mountain ranges and how they form. Seeing the Voyager pictures of Ganymede's bright terrain in 1979 really got me interested in planetary problems, in trying to apply what we know about terrestrial faulting to this icy surface. It's fascinated me ever since, and I've always wanted to get a closer look at what's going on there.
>
> I think of Ganymede as the Goldilocks moon of the Jupiter system in many different respects. It's got enough going on to be interesting, but not so much going on to be interesting enough to kill you. Not too old, not too young, some radiation here, not too much there. It's close enough to Jupiter but not too close. It's got a little bit of tidal action but not too much in the way of tides. [From a geological standpoint] you have Europa and Io on the exciting end, and Callisto on the not-exciting-enough end, so Ganymede is perfect. It's got Europa-like aspects and Callisto-like aspects. So why go to Europa or Callisto when you could go to Ganymede? Europa has these strange bands that have pulled apart, these wedges. It looks like Ganymede has some of those, too. Ganymede even has some places where it looks like it has undergone Europa-like activity. We think it may have gone through this episode of a lot of tidal heating, a sudden burst of activity, and then it's been fairly quiet since then. So it's like seeing an old version of Europa.
>
> Jupiter is going to be a lot bigger in the sky [than at Callisto], a lot more impressive. That magnetic field helps you out, too. There are certain parts of Ganymede that might be more like, say, being in the Colorado mountains; you'd feel like you were surrounded by lots of mountain ranges. The skiing possibilities are much better than on Callisto. Europa would be fascinating, but you'd get cancer pretty quickly. Io would be spectacular, briefly, and then you'd be dead.
>
> As far as a tourist destination, Ganymede also has the aspect of being easier to get into orbit. It's the largest mass in the Jupiter system. You can use it as a good gravitational anchor, both from which to explore the rest of the Jupiter system but also in terms of getting stuff down to the surface and getting into orbit around Ganymede. It's easier than getting into orbit around Europa. Callisto is similar; it's a little smaller and it has some of the same advantages. But it's farther out. It doesn't have as good a view of Jupiter.
>
> NASA was having a competition between different flagship mission concepts, and I was on a team that was studying a mission that was supposed to look at the whole Jupiter system at once. The solution we arrived at was that the best place to observe everything that was happening in the Jupiter system would be to basically build a Hubble Space Telescope and park it around Ganymede. You could watch from an ideal, safe location protected from radiation and in a place easy to get into orbit. So if we wanted to explore the Jupiter system, Ganymede is a great home base to do that from. It has many advantages in that respect. Europa and Io are close enough to look great. And if you want to watch Io volcanoes from a safe distance, you want to go to Ganymede.

CALLISTO

Callisto completes our tour of the Galilean satellites in the outermost orbit of the four. An astronaut standing on its anti-Jovian side would gaze out into the emptiness of space, with no hint that the largest planet in the Solar

System loomed somewhere below the horizon. Star-like moons would sometimes break the starry monotony. But on the sub-Jovian side, the story would be quite different. A viewpoint near the equator, but toward the edge of the anti-Jovian hemisphere, would yield spectacular views of the king of worlds, hovering at the horizon. Jupiter would seem to be plastered in one place, affixed to the crystalline celestial sphere. Like our own Moon, it would clock steadily through its phases from crescent to gibbous to full, and then back again in reverse.

Jupiter itself would span a section of the firmament as far across as nine full Moons in our own sky. The other Galilean satellites would trundle back and forth along a line across the equator of their giant primary. Ganymede's orbit would carry it out to an apparent distance of 65 full Earth Moons, while Europa and Io would stick closer to the planet, crossing it more quickly. Ganymede, closest to Callisto and largest of the four satellites, would appear to swell to the size of the Moon in terrestrial skies as it glided close by Callisto. Europa and Io would both appear about a fifth the size of Luna.

As morning comes to Callisto, Jupiter's full disk begins to wane. The Sun breaks over the horizon, casting long shadows from pinnacles of pure water-ice. Perhaps spectral tones would play across the landscape within those shadows, illuminating ancient, soft craters on mahogany plains of fine powder. The rising Sun illumines slopes on those plains where dusty landslides have left tongues of material fanning out from icy promontories. It's a beautiful, bleak place.

Callisto's battered face is geologically the oldest of the Galileans, with ancient craters probably dating back to the initial heavy bombardment, a rain of meteors and asteroids during the formative years of the Solar System some 3.8 billion years ago. Its ubiquitous craters point to a fairly quiescent internal structure. Callisto has no traces of Io's volcanoes, no hint of Europa's leaky ridge systems or fractured surface, none of the bright scoring on the moon next door, Ganymede. Callisto is a world unto itself.

The internal structure – or lack of it – also sets Callisto apart from the other Galileans. APL's Louise Prockter comments, "Ganymede was active far into its history compared to Callisto, which really didn't do anything; it didn't even differentiate properly." Differentiation, the process of heavy material sinking to create a dense metallic/rocky core, took place on all the other Galileans. But gravity studies show that Callisto is a muddled mix of ice and rock.

Callisto's odd interior contrasts with the other three Galileans, in part, because of a unique orbital interrelationship between Io, Europa and Ganymede. Each time that Ganymede circles Jupiter, Europa orbits twice; Io carries out a quartet of revolutions in the same time span. This strange relationship means the three moons are in 'resonance'; they are linked to each other in a gravitational dance that heats their interiors. This resonance has, in effect, kept the three siblings young at heart, heating their interiors and triggering active geology on their surfaces.

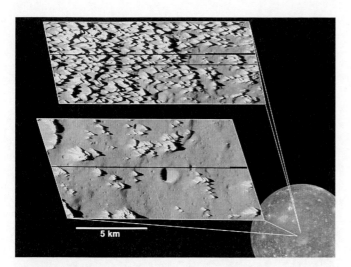

Fig. 5.23 Pinnacles and mesas of bright ice tower above Callisto's dark brown regolith. The surface material, the consistency of fine powder, appears to be left behind as the moon's icy surface sublimates. This photo was taken in the central region of the vast Asgaard impact basin (Image courtesy of NASA/JPL)

However, if every family has a dysfunctional member, Callisto fills the role for the Galilean clan. In the active geology department, Callisto is an underachiever. Its battered surface bears witness to a quiet landscape virtually unchanged from within, sculpted by a drizzle of meteoroids, comets and asteroids since its surface first solidified. In the Voyager era, Callisto was seen as the most boring of all the Galilean satellites – no volcanoes, no geysers, no shifting ice crust or folding mountains. But the Galileo spacecraft revealed a more complex portrait of the giant moon.

The undulating horizon of Callisto is pierced by what may be the most remarkable aspect of its landscape – great pinnacles and knobs of water-ice that rear up from its chocolate-brown plains. "The landscape has disintegrated," says William McKinnon, "as if something has been chewing on the crater rims."

Ames Research Center's Jeffrey Moore has been studying the frozen Callisto wonderland and, in particular, those dramatic frozen protrusions and the processes that led to them. "They seem to most often form from degraded crater rims; that's the major relief-forming geologic process on Callisto. The craters themselves are made up of bedrock, but they seem to have a lot of volatiles in them, probably dry ice [frozen carbon dioxide]. Once you expose the bedrock, which is a mixture of water-ice, dry ice and a fine powder of refractory material.[10] the dry ice decays away, causing the traps to retreat. The water-ice remains in the local cold traps, crater rims and things like that." The result is a series of several hundred meter-high ice pinnacles. "You grow a landscape that whimsically does not look much different from Monument Valley," Moore adds, "where the pinnacle part of it looks like bright, shining frosty ice."

In places, the pinnacle-making process leaves behind a refractory lag of fluffy dust. The lag blankets the surface, suppressing sublimation and protecting the ice beneath it. The dark stuff left over from the bedrock has the thermal properties of talcum powder. "You might be standing on a vast plain of fluffy, dusty dark gray material looking up at all these buttes and pinnacles that are these big, shining towers of ice. So what seemed like the most boring of all the Galilean satellites from a distance might easily be the most spectacular one to walk around on and visit as a tourist."

Jeffrey Moore is a planetary scientist at NASA's Ames Research Center in Mountain View, California. His work centers on the geologic evolution of planetary landscapes. Lately, Dr. Moore has concentrated on the icy satellites:

The icy satellites fascinated me because they were exotic landscapes. I'm a geologist and I like places that are interesting and mind-boggling. I'd just gotten out of the army in the late 1970s just as Voyagers 1 and 2 flew past the Galilean satellites.

10. Material not volatile enough to melt away as the ice does. Rather, it says behind as the CO$_2$ and water-ice evaporate.

First and foremost, they were spectacularly different from each other as well as anything we'd seen in the inner Solar System, so they were the hot cool new thing when I was an undergraduate. So in my undergrad and graduate years the Voyagers were showing us these icy satellites. I started doing work on those along with working with what was then not-very-old Viking data on Mars.

It's clear that the rocket barons of the twenty-first century think they can make money off space tourism. So if space tourism seems to be a possibility, I think it's a no-brainer: Callisto. You'd build a few tourist resorts with a spectacular view of Jupiter, with the Monument Valley-like foreground around you, like visiting Moab. Those scenes would be pretty hard to beat, but you'd also get great views of the other Galilean satellites and Jupiter itself. Callisto is it for the 'wow' factor.

The other advantage of Callisto is it's the only Galilean satellite you can visit and not get properly fried by the radiation. It's outside of Jupiter's radiation belts. It's such a nasty environment to actually send humans out to Europa, which is certainly one of the most interesting satellites scientifically, if you are interested in life in the Solar System (which most people are). Ganymede is not as bad as Europa or Io; you might get by in your lead underwear. You might still put your human base in or around Callisto, and simply tele-operate your equipment on the surface of Europa or Io. The time lag would be somewhere between four and eight seconds [depending on where you are in your orbit]. It's enough to be annoying, but it sure beats the heck out of a 45-minute time lag from Earth.

Fig. 5.24 Callisto (JPL)

Perhaps the exotic Galileans will prove a great temptation to those future "rocket barons" Moore refers to. Undoubtedly some braver members of the scientific community will venture out to investigate them in situ. In time, the strange moons first spied by Galileo and Maurius may become the next playground of the rich, a cosmic equivalent of the Caribbean or Canary Islands. And though there will be no palm trees, the Galileans offer a rich variety of seas and a varied host of wild landscapes in which to revel.

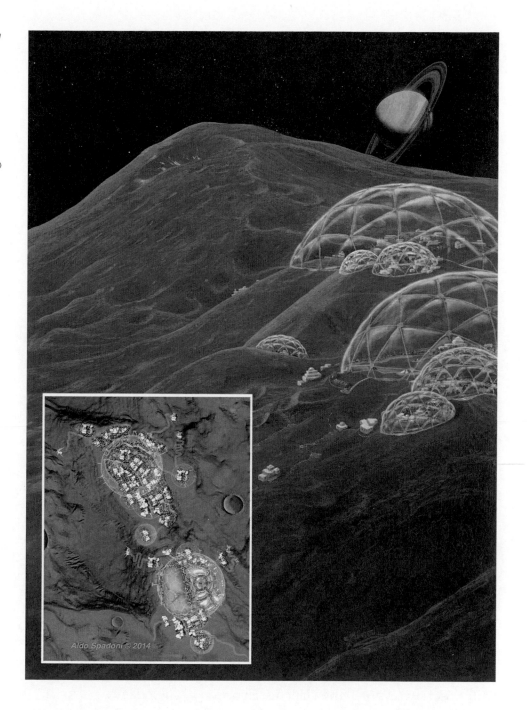

Fig. 6.1 The Ridge Resort – an all-inclusive vacation facility perched on the equatorial ridge of Iapetus would witness spectacular views of Saturn and its rings. Sites like this would have Earthly analogs in tourist spots like the cliff-hugging villages on Greece's Santorini (Painting © Michael Carroll; city design and inset © Aldo Spadoni. Used with permission)

Chapter 6

Saturn's Ice Moons: Dione, Tethys, Rhea, Hyperion, Iapetus, Phoebe and Enceladus

The moons of Saturn provide many inspiring destinations for future explorers. Each has its own set of challenges for landing and traveling on the surface, and each affords a unique view of nearby Saturn, its spectacular butterscotch clouds, glowing aurorae and vast rings.

Beyond Jupiter, the Lord of the Rings coasts majestically around the Sun once each 29½ years. The golden orb is encircled by the most extensive rings, and the most extensive family of satellites, of any planet in our system. Of its 62 confirmed major moons,[1] 13 are larger than 50 km/ in diameter (see Fig. 6.7). With the exception of planet-sized Titan, these moons make up a mid-sized family of bodies that are more poorly understood than the larger Galilean satellites.

Moons of this size tend to perplex researchers, says APL's Elizabeth Turtle. "We have a much better understanding of how larger satellites work, especially the Galilean satellites. We have a history based on those satellites, which form a nice sequence between end members. But the mid-sized satellites aren't terribly well understood." Turtle points out that Saturn's largest mid-sized moon, Rhea, is less than half the diameter of the smallest Galilean, Europa.

BATTERED ICE WORLDS: DIONE, TETHYS, RHEA AND HYPERION

Jean-Dominique Cassini discovered four of Saturn's moons in the late seventeenth century. He wanted to name them the *Sidera Lodoicea*, the "Stars of Louis," to honor France's King Louis XIV. But the moons ended up with names of the Titans, the sisters and brothers of Kronos (the Greek counterpart of Saturn).[2] They were the moons Dione, Tethys, Iapetus and Rhea.[3]

Saturn's moons seem to come in pairs. Moving out from the planet, small Mimas and Enceladus both subtend a diameter of about 500 km. Tethys and Dione span diameters of a 1,000 km, while beyond them Rhea and Iapetus measure about 1,500 km across. When *Voyager 1* and *2* blazed through the Saturnian system, humankind got its first clear look at the ice worlds. Each spacecraft carried out a different assignment: *Voyager 1* targeted detailed studies of Titan, Dione, Rhea and Mimas. Flight engineers tasked *Voyager 2* with closer surveys of Tethys, Iapetus, Hyperion and Enceladus.

The environs of Saturn dictate forms on the surfaces of its ice moons. Average daily temperatures rise up to a scant −187 ° C. At those cryogenic temperatures, the water-ice crust of these small satellites behaves much like granite. Researchers anticipated that Saturn's family of mid-sized ice worldlets would present unique features due to their compositions and temperatures. As the Voyagers approached, investigators waited to see just what kinds of landscapes would result from those temperatures on icy substances.

1. As of this writing, the Cassini mission continues to discover moons on a fairly regular basis, most of which are tiny (less than 500 m/0.3 miles in diameter). In the dense B-ring alone, 150 have been charted. These moons are small enough that they cannot clear a consistent gap, like the Keeler gap or the Cassini division. Rather, they clear small, propeller-shaped areas in front of and behind themselves.

2. British scientist Sir John Herschel suggested the naming convention. He was the son of William Herschel, who discovered the planet Uranus. The elder Herschel also discovered two moons of Saturn, Mimas and Enceladus. In keeping with the family tradition of cosmic discoveries, William's sister Caroline not only served as his assistant but discovered several comets and nebulae in her own right.

3. By this time Titan was already known, having been discovered by Christian Huygens in 1655.

M. Carroll, *Living Among Giants: Exploring and Settling the Outer Solar System*, DOI 10.1007/978-3-319-10674-8_6, © Springer International Publishing Switzerland 2015

Fig. 6.2 Dione passes in front of Tethys in this snapshot from the Cassini orbiter (Image courtesy of NASA/JPL/SSI)

Voyager 1 imaged Dione at short range, revealing cratered plains etched by several wandering fractures. *Voyager 2* imaged other territory not seen by its sister craft, but at a further distance. *Voyager 2's* distant images covered the other side of the moon and displayed mysterious wispy bright lines, like mist plastered to the surface. Dione has differing leading and trailing hemispheres. The feathery bright markings meander across the dark trailing hemisphere, while the leading hemisphere is brighter and more heavily cratered. At first, some investigators theorized that early in its history, Dione's surface had been brightened by cryovolcanic eruptions of fresh ice along fractures. After the geologic activity died out, the theory went, cratering on the leading hemisphere obliterated evidence on that face, leaving the wisps only on the relatively sheltered trailing side.

These regions remained a mystery until the Cassini spacecraft settled into orbit around Saturn for its multi-year reconnaissance. Its repeated close encounters of Dione solved the riddle of the white wisps. They are actually steep ice cliffs rising from the cratered plains as great walls hundreds of meters high. The cliffs cut across craters, indicating that they formed after the earlier heavy bombardment of the moon.

"Dione is distinctive," says John Spencer, an expert on small icy bodies of the Solar System at the Southwest Research Institute in Boulder, Colorado. "A lot of the surface is smooth and seems younger than the other moons. The craters have a distinctive shape, as if there's been enough heating to flatten them out. It's the next to Enceladus in terms of activity, far below Enceladus, but it's above Tethys and Rhea and Iapetus."

Dione's mass indicates a large amount of rock mixed with its ice, perhaps enough to cause radiogenic heating early in its life. This heating, along with gravitational pulling from other moons, might have triggered geyser-like activity in the past. Today, Dione orbits Saturn in a 2:1 resonance with Enceladus, helping to tug at Enceladus' interior and generate geyser activity. At one time, Dione may have been in resonance with other moons in such a way as to cause energetic heating of its interior. But Dione itself seems a quiet place today.

Appearances may be deceiving, according to a recent article in the journal *Icarus*.[4] Observers studied the large mountain Janiculum Dorsa, an 800-km-long ridge that in some places towers nearly 2 km above its cratered surroundings. It is those surroundings that give hints about the

---

4. *Icarus*, March 2014.

interior of the moon. The landscape around the mountain seems to have sunk under its weight, implying that a subsurface ocean of liquid water or slush lies below. The depressed surface proves that, at the very least, the crust was at one time warmer, and its heat must have come from the ice moon's interior. Is it possible that, at one time, this interior erupted out in geysers or icy volcanoes? JPL planetary scientist Bonnie Buratti says Dione may be "a fossil of the wondrous activity Cassini discovered spraying from Saturn's geyser moon Enceladus." It is also possible that the moon is still geologically active, with an interior liquid ocean. The Cassini space-craft's magnetometer sensed a weak stream of particles drifting away from the moon, which might imply one. Dione also has a rarified atmosphere – called an exosphere – of nearly pure oxygen atoms.

Fig. 6.3 Voyager 2's distant views of Dione (upper left) provided tantalizing views of wispy terrain. Over two decades later, the Cassini Saturn orbiter revealed the wisps as ice walls thrust up by faulting (upper center and right). This detail of the rugged landscape (bottom) shows a crater at the center that has been sliced by one of Dione's great fractures (Upper left image courtesy of NASA/ JPL; all other images courtesy of NASA, JPL, Space Science Institute)

If liquid seas do exist beneath Dione's crust, the moon would join Europa, Ganymede, Callisto, Enceladus and perhaps Titan as a host to an alien ocean of water, increasing its importance as a destination for astrobiologists and other explorers.

The next two moons out, Tethys and the much larger Rhea, also wear cratered countenances. But glistening Tethys is much less dense than its adjacent siblings, made up of nearly pure water-ice. It is the brightest of all Saturnian moons except Enceladus. Its surface is less heavily cratered than Dione and Rhea, suggesting that internal heat kept its surface malleable until after most early craters formed. This internal heating may have come

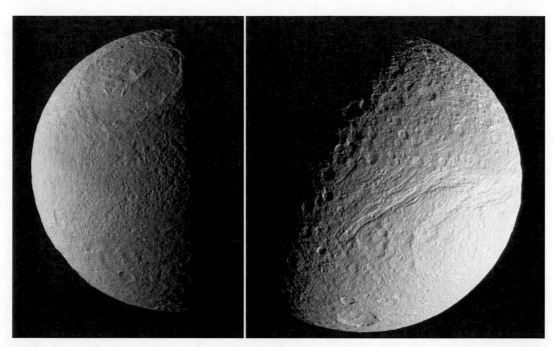

Fig. 6.4 *The crater Odysseus (left) and the canyonlands of Ithaca Chasma are features unique to Tethys (Image courtesy of NASA/JPL, Space Science Institute)*

from a resonance with other moons, although Tethys is free from such gravitational relationships today.

Tethys has two dramatic geologic features. At 445 km across, the massive crater Odysseus spans 2/5 of the diameter of Tethys itself. The impact must have been nearly violent enough to destroy the moon. A crater similar in relative size to its parent body was spotted in Voyager images of tiny Mimas. But unlike Mimas, Tethys is large enough that the floor of its mega-crater filled in with icy slag after the impact, taking on the same curve as the rest of the moon's surface. This suggests that Tethys had a warm interior at the time of the event. If Tethys had been solid ice, the impact might well have fragmented the moon into a new Saturnian ring.

The second remarkable feature on Tethys is a complex of canyons called Ithaca Chasma. Stretching nearly from pole to pole, the abyss lies almost directly opposite the crater Odysseus. Ithaca may well have opened up in the crust as shock waves from the Odysseus impact moved through the moon. It is also possible that the canyon is a scar left from the expansion of the moon as its water-ice froze and expanded.

The third satellite of these mid-sized moons is Rhea, largest of Saturn's frozen companions, and second in size only to Titan. Rhea is one of the most blasted moons in the Solar System. It has far more craters per square mile of real estate than the other mid-sized Saturnian satellites. Many of its regions exhibit saturation cratering (craters so dense that when new craters form, they obliterate the same number of older ones). A preponderance of craters less than 20 km across pepper other regions.

This second population of impacts may have come from within the Saturnian system itself, blown from the surfaces of nearby moons. One candidate is the outer moon Hyperion, as we will see.

Rhea is denser than Tethys, and probably consists of about three-quarters ice and one quarter rock. Like Callisto, Rhea seems to lack a differentiated core and mantle, with a jumbled mix of rock and ice in its interior. And like Dione, Rhea has dissimilarities between its leading and trailing hemispheres. Areas at the poles and equator have fewer large craters than other regions, hinting at some resurfacing event in Rhea's past. Like Dione, Rhea's trailing edge has several bright cliffs. In Rhea's case, these crags form canyon walls.

Rhea's size and mass may actually prevent it from experiencing significant geological activity. The moon is not in resonance with any other moon, so any surface changes from internal forces would rely on energy from radiogenic heating. Since there is no core to collect and preserve all that heat, it must have been evenly distributed throughout; Rhea must have cooled gradually. As it did, interior ice became more dense and the globe shrank, compressing the surface and shutting down any early cryovolcanism. Like the wrinkles on a raisin, Rhea's ridges and cliffs may be the result of such a process.

Rhea also sports a tenuous exosphere of oxygen and carbon dioxide. Its source is unknown, but the gases may either escape from the interior or emanate from surface ices as they are bombarded by radiation.

The fourth of the battered worldlets is the strange moon Hyperion. Hyperion is easily large enough to be round, measuring just less than Mimas at its longest. But the bizarre little ice chunk is quite irregular. In addition to its shape, the moon is not tidally locked. Any irregular moon should point its longest axis toward its parent planet, but Hyperion somersaults dramatically as it circles Saturn. Its tumbling motion is essentially chaotic, something unseen on other bodies of its size.[5] These anomalies suggest that the moon has been chipped away by violent impacts. Its cratered surface looks like a spongy battlefield, and its low density indicates that the spongy appearance is more than skin deep; the ice moon is remarkably porous. Says planetary geologist Jeff Moore, "Hyperion looks like it's had the crap smashed out of it, so it's the poster child for what a mid-sized satellite should look like."

Hyperion's craters are unlike craters on other moons. Its impact scars seem to have melted, expanding the craters into deep pits. The jagged-edged hollows lie cheek-to-jowl, some merging into others. Researchers conclude that Hyperion's low density – just half that of

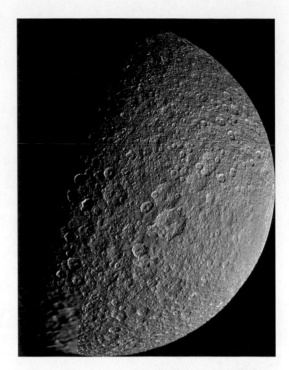

Fig. 6.5 Rhea's pummeled surface bears witness to a hail of asteroids, meteors and comets (Image courtesy of NASA/JPL, Space Science Institute)

5. Hyperion is in a 3:4 resonance with Titan, which may also contribute to its wild dance around Saturn.

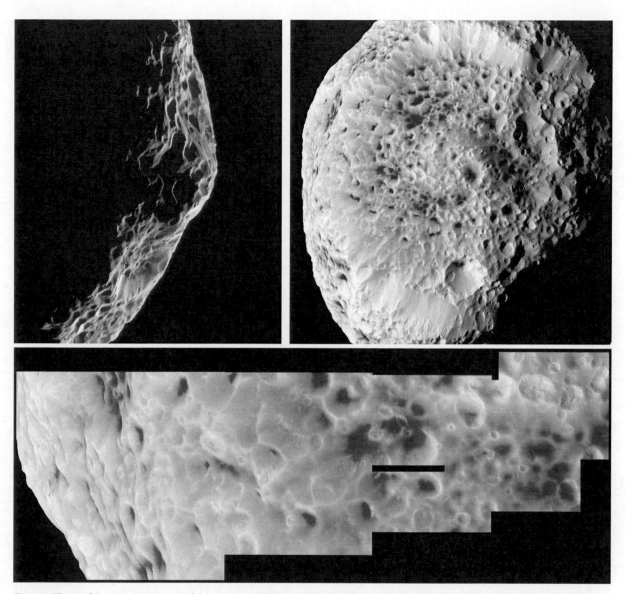

Fig. 6.6 Three of the Cassini spacecraft's best views of the tumbling moon Hyperion (Image courtesy of NASA/JPL, Space Science Institute)

water – and low gravity react in unique ways to impacts. Infalling cosmic debris tends to compress the little moon's surface, rather than exhuming it. Instead of forming secondary craters, most of the material from impacts gusts outward and never returns to the surface. An unknown material fills the interiors of many pits, and landslides are evident on some crater walls.

The Known Satellites of Saturn

	Satellite	Diameter (km)	Distance from Saturn (x 000km)	Discoverer	Date
-I-	S2009 S1	~.3	117	Cassini/Huygens mission	2009
-I-	Ring moonlets	~.04 - .4	~130	Cassini/Huygens mission	2006
1.	Pan	34x31x20	133.6	M. Showalter	1990
2.	Daphnis	9x8x6	136.5	Cassini/Huygens mission	2005
3.	Atlas	41x35x19	137.7	Voyager 2	1980
4.	Prometheus	136x 79x 59	139.4	Voyager 2	1980
5.	Pandora	104x 81x 64	141.7	Voyager 2	1980
6.	Epimetheus^	130x 114x 106	151.4	Fountain, Larson	1977
7.	Janus^	203x 185x 153	151.5	Dolfus	1966
8.	Aegaeon	~.5	167.5	Cassini/Huygens mission	2008
9.	Mimas	397	185.4	Herschel	1789
10.	Methone	~3.8	194.4	Cassini/Huygens mission	2004
11.	Anthe	~1	197.7	Cassini/Huygens mission	2007
12.	Pallene	5	212.3	Cassini/Huygens mission	2004
13.	Enceladus	504.5	237.9	Herschel	1789
14.	Tethys	1062.5	294.6	Giovanni Cassini	1684
15.	Telesto	33x 24x 20	294.6	Smith, Reitsema, Larson, Fountain	1980
16.	Calypso	30x 23x 14	294.6	Pascu, Seidelmann, Baum, Currie	1980
17.	Dione	1123	377.4	Giovanni Cassini	1684
18.	Helene	43x 28x 26	377.4	Laques and Lecacheux	1980
19.	Polydeuces	~.03	377.4	Cassini/Huygens mission	2004
20.	Rhea	1527	527	Giovanni Cassini	1684
21.	Titan	5151	1222	Huygens	1655
22.	Hyperion	360x 266x 205	1481	Bond, Bond and Lassell	1848
23.	Iapetus	1469	3561	Giovanni Cassini	1684
24.	Kiviuq	~16	11295	Gladman, Kavelaars et al.	2000
25.	Ijiraq	~12	11355	Gladman, Kavelaars et al.	2000
26.	Phoebe	219x 217x 204	12870	Pickering	1899
27.	Paaliaq	~22	15103	Gladman, Kavelaars et al.	2000
28.	Skathi	~8	15673	Gladman, Kavelaars et al.	2000
29.	Albiorix	~32	16267	Holman	2000
30.	S/2007 S2	~6	16560	Sheppard, Jewitt, Kleyna, Marsden	2007
31.	Bebhionn	~6	17154	Sheppard, Jewitt, Kleyna, Marsden	2004
32.	Erriapus	~10	17237	Gladman, Kavelaars et al.	2000
33.	Skoll	~6	17474	Sheppard, Jewitt, Kleyna	2006
34.	Siarnaq	~40	17777	Gladman, Kavelaars et al.	2000
35.	Tarqeq	~7	17911	Sheppard, Jewitt, Kleyna	2007
36.	S/2004 S13	~6	18056	Sheppard, Jewitt, Kleyna	2004
37.	Greip	~6	18066	Sheppard, Jewitt, Kleyna	2006
38.	Hyrrokkin	~8	18168	Sheppard, Jewitt, Kleyna	2006
39.	Jarnsaxa	~6	18557	Sheppard, Jewitt, Kleyna	2006
40.	Tarvos	~15	18563	Gladman, Kavelaars et al.	2000
41.	Mundilfari	~7	18726	Gladman, Kavelaars et al.	2000
42.	S/2006 S1	~6	18930	Sheppard, Jewitt, Kleyna	2006
43.	S/2004 S17	~4	19099	Sheppard, Jewitt, Kleyna	2006
44.	Bergelmir	~6	19104	Sheppard, Jewitt, Kleyna	2004
45.	Narvi	~7	19395	Sheppard, Jewitt, Kleyna	2003
46.	Suttungr	~7	19579	Gladman, Kavelaars et al.	2000
47.	Hati	~6	19709	Sheppard, Jewitt, Kleyna	2004
48.	S/2004 S12	~5	19906	Sheppard, Jewitt, Kleyna	2004
49.	Farbauti	~5	19985	Sheppard, Jewitt, Kleyna	2004
50.	Thrymr	~7	20278	Gladman, Kavelaars et al.	2000
51.	Aegir	~6	20483	Sheppard, Jewitt, Kleyna	2004
52.	S/2007 S3	~6	20519	Sheppard, Jewitt, Kleyna	2007
53.	Bestla	~7	20570	Sheppard, Jewitt, Kleyna	2004
54.	S/2004 S7	~6	20577	Sheppard, Jewitt, Kleyna	2004
55.	S/2006 S3	~6	21076	Sheppard, Jewitt, Kleyna	2006
56.	Fenrir	~4	21931	Sheppard, Jewitt, Kleyna	2004
57.	Surtur	~6	22289	Sheppard, Jewitt, Kleyna	2006
58.	Kari	~7	22321	Sheppard, Jewitt, Kleyna	2006
59.	Ymir	~18	22430	Gladman, Kavelaars et al.	2000
60.	Loge	~6	22984	Sheppard, Jewitt, Kleyna	2006
61.	Fornjot	~6	24505	Sheppard, Jewitt, Kleyna	2004
62.	S/2004 S6*	~3-5			2004
63.	S/2004 S3 and S4*	~3-5			2004

-I- The International Astronomical Union has yet to determine if small bodies like these should be counted as moons.
^Janus and Epimetheus are co-orbital satellites, meaning they share essentially the same orbit, exchanging places every four Earth years.
*These moons, near the F ring, are not confirmed.

THE OUTLIERS, IAPETUS AND PHOEBE

Iapetus

Since the days when Jean-Dominique Cassini discovered it in 1671, observers knew there was something different about Saturn's moon Iapetus. Cassini first discovered the moon to the west of Saturn, but when he tried to find it on its eastern trek, the moon seemed to have disappeared.

Years later, armed with more powerful instruments, the dedicated astronomer finally spotted Iapetus on the other side of its orbit, appearing far dimmer than it had when it was to the west. Cassini rightly concluded two things. First, Iapetus was much darker on one side than the other, and second, it was tidally locked to Saturn, keeping the same face always to the Ringed Planet. Who wouldn't? The view from Iapetus would have been spectacular!

The orbit of Iapetus is the most inclined of the regular satellites; only the irregular outer moons such as Phoebe and the smaller satellites have more inclined orbits. Iapetus is the most distant of the regular mid-sized moons of Saturn. Although it is the planet's third-largest moon, Iapetus orbits much further from Saturn than the next major moon in, Titan.

Iapetus' trailing hemisphere is as bright as dirty snow, while its leading side is nearly as dark as asphalt. This dichotomy baffled astronomers for centuries, up until the Space Age. Some proposed that internal volcanism was resurfacing a naturally bright moon. Others felt that infalling material might be peppering the dark hemisphere from above, perhaps from Saturn's dark outer moon Phoebe. But even the best telescopes on Earth could not solve the mystery of the "yin-yang" moon.

Fig. 6.8 Iapetus has dramatically different leading (left) and trailing hemispheres (Image courtesy of NASA/JPL/ Space Science Institute)

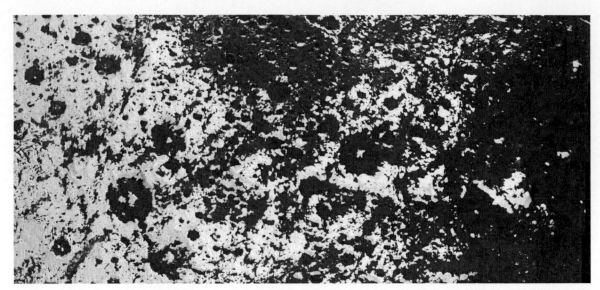

Fig. 6.9 The transition between dark terrain and light is, in places, abrupt (Image courtesy of NASA/JPL/ Space Science Institute)

In 1981, *Voyager 2* flew within 720,000 km of the strange moon, and the Cassini orbiter has since had several encounters at even closer range. Spacecraft images reveal that both the dark and light hemispheres of Iapetus are heavily cratered. At first glance, the material did seem to be welling up inside some craters, but further study hinted at a rain of material falling onto the leading side of the moon. Adding to the mystery was that the dark areas were far more reddish than the prime suspect for the rain of dust, Phoebe. The boundary between hemispheres is far too sharply defined to have been caused by simple dark "snowfall."

Clues to the true nature of the darkened region come from the boundary between it and the lighter ice. In this transition zone, in the highest resolution images, the scattered pools and curtains of dark materials fill in valleys and craters. Cassini radar shows that it is a very thin layer, in many places less than a meter/yard deep. Some small impacts have pierced through to bright ice beneath. Scientists have come to the conclusion that a complex set of processes contribute to the piebald nature of the moon.

One process at work is called thermal segregation. Water-ice migrates from illuminated, warmer areas, like Sun-facing crater walls, to nearby shadowed areas that are colder. The ice leaves behind a lag of darker material, while brightening the shadowed terrain.

Tilmann Denk of the German Aerospace Center's Institute for Planetary Research has been studying the ice migration on Iapetus. He suspects that the dark material, whether intrinsic to Iapetus' surface or deposited from the outer moons, is powdery. "Thermal data from Cassini's CIRS instrument implies a fluffy surface. This means that an astronaut might sink more than he or she would do on the Moon, unless some kind of sintering or other compaction process acted just below the surfaces. But the astronaut would not disappear in the dust, maybe just sink down to the knees or so, as a guess."

Infalling dust does contribute to the darkened terrain. It appears to come from Phoebe, orbiting outside the orbit of Iapetus. In fact, Cassini team members discovered a vast ring of dust encircling Saturn and coming from Phoebe. This dust tends to drop toward Saturn, and the moons in between get in the way. The dust is swept onto their surfaces, and especially the surface of nearby Iapetus. Over time, the dark dust heats the surface ice, warming it near the equator enough that it evaporates. That ice condenses again at the bright poles and around on the trailing hemisphere. The low gravity of Iapetus – only 20 % of the ice moon is rock – enables ice to move freely around the globe, creating one of the most remarkable sites in the Solar System.

But Iapetus has another wonder – a titanic ridge that stretches along the equator like the seam on a cheap rubber ball. The ridge towers some 13 km high and spreads as wide as 20 km. The peaks making up portions of the ridge are among the tallest mountains found on any planet or moon. The great ridge branches into three parallel crests on one end. Some of its segments are 200 km long, and other subdivisions break up into isolated peaks. The spectacular rise is heavily cratered, implying that it formed early in Iapetus' history. The ridge is one of the most perplexing features in the entire Solar System. It shares three elements that make it unique. It sits exactly on the equator, it is limited to the equatorial region and nowhere else and nothing like it is found on any other planet or moon.

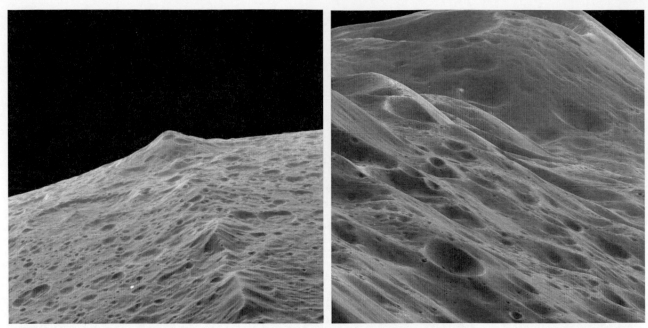

Fig. 6.11 The mysterious equatorial ridge of Iapetus has no known equal among the worlds of our Solar System (Image courtesy of NASA/JPL/ Space Science Institute)

Some postulate that the ridge resulted from a slowing of Iapetus' rotation. But why did the ridge form only on the equator, unlike the tectonic features found all over Europa and Ganymede due to similar forces? John Spencer has struggled with the issue. "The one thing that's distinctive about Iapetus from any other moon in the Solar System is that it's rotating synchronously, but it's very far from its primary object, farther out than other synchronously rotating objects. Things rotate synchronously because of the tides of the parent body, just like our own Moon does, and those tides are much weaker on Iapetus. Initially, it seemed reasonable to me that if the de-spinning would be happening so late it would happen very slowly, it would have taken hundreds of millions of years, and maybe you could still see the effects of this de-spinning which would have been erased on the other moons. That might cause forces on the equator that would buckle up this ridge. But the geophysicists say, 'No, it doesn't work, you would see more complex patterns than you do see.' You would get a lot of deformation as it went from an oblate shape to a spherical shape, but it would be very hard to focus those in such a way that all you get is this ridge at the equator. I am persuaded that was probably not what was going on."

A similar concept proposes that as it formed, Iapetus was spinning so rapidly that centrifugal force pushed it up at the equator. As a planet or moon forms, it "spins up" as it condenses out of a cloud of material. Presumably Iapetus was spinning much faster when it froze into its current shape. But a body as large as Iapetus would tend to become oblong under such forces, and if it froze solid when spinning with a 17 h period, it would still be spinning close to that rate today. A frozen-solid Iapetus would not have enough friction in its interior to brake its rotation to today's Saturn-synchronous turn.

Others propose that the ridge was thrust up from forces below, but this would require Iapetus' stiff outer layer to be relatively thin. In fact, the moon's crust is thick enough that it holds up those tall mountains without deforming around them, as it would were it thinner. The sinking of the crust around a heavy formation is called flexural deformation. This phenomenon can be seen on all the terrestrial planets and many moons that have solid crusts.

Whether by shifting temperatures, fluctuating air pressures, or the flow of liquid to the lowest point, natural forces constantly battle to reach equilibrium. This equilibrium, or balance, is called isostasy. On Earth, thickening of the crust (for example, a mountain) weighs heavily on the fluid mantle below. The mantle shifts away from the pressure above it, letting the heavy object settle into isostasy. The crust around the object bends inward toward it.

On our own world, this "moat" around the mountain or other structure is quickly filled in with sediment, so it is difficult to see (these depressions can be seen in gravity maps). But on a moon such as Iapetus, if the mountain rises up atop a fluid interior, the ground around it should obviously sag. No such moat surrounds the Iapetus ridge, leading analysts to conclude that it did not emerge from the interior but rather was somehow deposited onto a solid, thick crust.

William McKinnon and a team of planetary geologists[6] recently proposed a theory that explains all of the ridge's unique characteristics. Analyzing computer models and Cassini's most detailed images, the team describes a scenario in which a debris ring around Iapetus – the leftovers of a small moon – gradually collapsed on the equatorial zone.

Any small moon captured by a larger body settles into an elliptical orbit, but over time tidal forces would circularize its path. Depending on its direction of travel, the moon would either work its way out further or spiral closer to Iapetus.

At the same time, Saturn's gravity would be slowing Iapetus' rapid spin, which would be fairly speedy after its initial creation. Eventually Iapetus becomes tidally locked, turning once for each time it travels around Saturn in its 79-day orbit. As Iapetus slows, it also slows the orbit of its own small moon, which begins to fall toward it. Soon, the gravity of Iapetus pulls the moon apart, scattering it into a thick ring of boulders, gravel, sand and dust. Eventually, the ring flattens, aligned to the equator just as Saturn's is aligned to its own. Because of Iapetus' low gravity and the dynamics of such a ring, particles would come down at glancing blows and very low speed. The rain of stony debris would build up material rather than excavating craters and destroying the surface.

Why have we not seen this on other moons? The reason is that most moons orbit near other ones, so gravity disturbs their environment, says John Spencer. "A ring around most moons would not be stable because of the tidal forces from the planet. But Iapetus is so far away from Saturn that you could have a longer-lived ring that wouldn't get ripped apart from

6. Dombard et al., Abstract P31D-01 presented at 2010 Fall Mtg., AGU.

those tidal forces." Because of its remote location, its gravity has an uninterrupted influence over large areas of space around it. In fact, its gravity rules over ten times as much space as nearby Titan. This vast influence may have enabled Iapetus to snag a passing asteroid and keep it as a moon that would one day become its exotic equatorial ridge.

Some researchers are skeptical of the ring debris scenario. They assert that the great crest shows signs of being uplifted from forces below. The formation certainly presents a conundrum, says Tilmann Denk.

> Some of my smart colleagues favor the deposition idea, while other smart colleagues favor internal geology. Both sides – and this is really interesting – have good arguments [explaining] why the other idea cannot be true, resulting in the situation that the ridge cannot exist. Remember that you do not just have to explain the existence of the ridge, but also why it spans only about one half of Iapetus, and why the shape of the individual mountains differs so much, from tent-shaped to trapezoidal cross-section with sharp edges, to individual mountains with smooth edges and a more triangular cross-section. Most obvious, still, is the absence of the ridge inside the huge Turgis Basin; this implies that the ridge is older than Turgis, or, in other words, really old.

All of the theories of ridge formation seem exotic to researchers like John Spencer. "The idea that there was a ring that collapsed onto the surface – which seems crazy, too – may be more likely. I still think it's crazy, but I don't have any better ideas."

Recently, observers detected a debris ring around the asteroid 10199 Chariklo, a body orbiting between Saturn and Uranus. This Centaur-class asteroid probably originated as a Kuiper Belt Object. It is estimated to be 250–300 km across, or one-third the size of Iapetus. William McKinnon comments, "What I think hasn't really been emphasized in the 'ring around a Kuiper Belt Object' discovery is that this is the first evidence for a ring around *any solid body* in the Solar System. In that sense it is a proof of concept that satellites could have satellites, or rings, at least for a while."[7]

7. For a good summary, see "Icy rings found around tiny space rock," *Science News*, May 3, 2014, p. 10.

Fig. 6.12 Landslide on Iapetus (NASA/JPL/SSI)

Clearly, the jury is still out on the origin of the great ridge, but if the debris theory is correct, we may see a similar formation on another moon that has a similar gravitational arrangement with its other siblings. That moon is Uranus' Oberon. It will be some time before we have high resolution data on this moon (see Chap. 3), so planetary researchers will have to bide their time, something they have become good at.

Tilmann Denk of the Free University of Berlin is a member of the Cassini imaging team. He helps to select targets for the spacecraft's imaging system among the icy moons of Saturn. Here is what he says concerning about being on Iapetus:

Iapetus' surface gravity is about 2 to 3 % of Earth's. I expect it would resemble the situation on the Moon with respect to keeping order with untethered objects like glasses or plates or laptops, but walking should be done with extreme care. This is certainly different than walking on Callisto. When doing a moderate jump on Earth (just vertical from standing, not a running jump), your center of mass might reach a height difference of ~0.2 m before falling down, and the fun is over after less than half a second. On Callisto, the same jump would reach about 3 m (quite more fun), with a return after 4 to 5 seconds. On Iapetus, it's ~20 m, with a return after about half a minute (no fun anymore?). A non-vertical jump on Iapetus might transport you ~40 m away from your starting position. A vertical jump with equipment about equal in weight to what an astronaut carries would still allow you an easy jump of ~5 to 10 m altitude or ~10 to 15 m wide. In other words, an "extravehicular activity" outside a base should not be done without securing devices or jet pack.

Up on top of the ridge, the difference would be tremendous. Standing on the flat ground on Iapetus allows for a horizon view around 1.5 km away. Standing on a 15-km tall mountain would increase this to ~150 km or ~11 degrees in longitude and latitude in viewing direction if there's no obstacle like another mountain. On Earth, a similar view can be obtained from about 1,800 m altitude.

However, on the Iapetus surface, there are spectacular avalanches. For example, the craters Engelier or Malun (see Fig. 6.11) contain nice examples. These might be touristically attractive, for example from the crater rim or by climbing. The view on the ridge or from the ridge would also be great. I can imagine a location where Saturn is just at the horizon (near the meridian running through the center of the leading or trailing side). This would mix the planet view with the mountainous landscape. Of course, you have to take care that libration, which causes a slight movement of Saturn in the sky, does not cause Saturn to sink below the horizon occasionally. Finally, the possibility of doing tremendous jumps is also a significant difference from a larger moon like Callisto. Some tourists might not like it, while others will consider this exciting.

Fascinating Phoebe

Well outside the orbit of Iapetus sails the last of the mid-sized satellites, Phoebe. The most distant of the major Saturnian satellites is an oddball. Phoebe breaks all the rules for a moon of its size. It goes around Saturn the wrong way, in a retrograde orbit. It is small enough that it should be a solid, irregular shape, but instead seems to be held in a nearly spherical form. And though it orbits some 13 million km away, Phoebe seems to be generating its own dust ring around Saturn.

Phoebe's orbit indicates that the moon winged its way in from afar. The moon circles Saturn in a retrograde course, opposite the direction that the Saturnian system naturally rotates. Its orbit is also eccentric – not circular – and inclined at an extreme angle. Phoebe orbits far from her parent world, circling the planet in a great loop some 27 million km across, taking a leisurely 18 months to make one circuit. These clues point to a capture of the moon early in the evolution of our Solar System. Phoebe came from somewhere else.

Phoebe's history is difficult to reconstruct, and offers many possibilities. If the moon came from the outer system, its travels from the Kuiper Belt may have taken it closer to the Sun, perhaps as close as Jupiter, before capture. Once in Saturn's vicinity, there may have been even more fireworks – Phoebe may have had a sister.

Dynamicists are skeptical that Phoebe could have been captured gravitationally if it came soaring into Saturn's neighborhood by itself. For Saturn to capture Phoebe, computer models indicate, a third object was necessary. In the encounter, Phoebe's companion would have been ejected, and the lost energy would slow Phoebe enough for capture.

Another intriguing possibility is that Phoebe was a moon of either Uranus or Neptune. In those wild days of planetary migration (see Chap. 2), one of those planets may have come close enough to Saturn to shed some of its own natural satellites, perhaps actually trading a few with the larger planet.

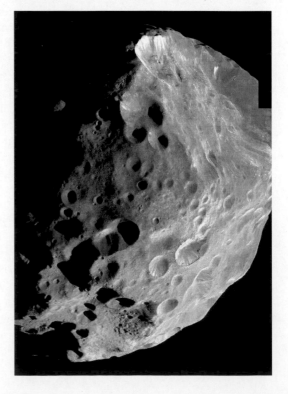

Fig. 6.13 Cone-shaped craters and a fairly-spherical shape make Saturn's distant moon Phoebe unique (Image courtesy of NASA/JPL/Space Science Institute)

As natural satellites go, the distant moon falls below the mid-range of familiar moons like Rhea or Mimas, with a diameter averaging 213 km. Other moons of similar size are not large enough for their weak gravity to pull them into a sphere. They are probably cold mountains of rock and ice. But Phoebe is spherical enough that internal heat may have softened it early in its history. Its density is higher than typical icy Saturnian satellites. This suggests that the tiny satellite may be differentiated by having a dense rocky core, putting it in the small club of planet-like bodies.

As we saw at Europa with the Galileo spacecraft, planetary interiors can be mapped using spacecraft flybys. When a spacecraft passes by a planet or moon, flight engineers chart the path of the craft as gravity bends its course. If the pass is close enough, or if there are enough passes, scientists can determine the internal structure of the passing body. Sadly, Cassini did not fly close enough to Phoebe for such calculations to be made. But the shape of Phoebe betrays characteristics of its interior. The moon's shape is close enough to round that it had to be partially molten in the past. When a planet relaxes into a spheroid, it reaches gravitational equilibrium; its shape is stable under the influence of its internal gravity. Phoebe seems to have reached this point of "global relaxation.".

Fig. 6.14 During its flyby of Phoebe on June 11, 2004, Cassini mapped the distribution of water-ice, ferric iron, carbon dioxide and an unidentified material on the tiny moon's surface (Image courtesy of NASA/JPL/ University of Arizona)

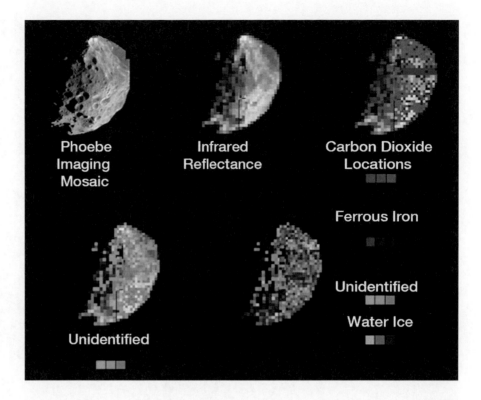

Phoebe's relaxed surface also affects the shape of its craters. Crater rims seem to be composed, in places, of pure ice. The craters themselves are oddly shaped, as if influenced by internal forces. Some look as if there were pockets of gas or volatiles that blew the craters out from the inside. Mass wasting (landslides) may also be to blame; several crater walls seem to have caved in. Many of Phoebe's craters are cone-shaped, unique among the icy satellites that have thus far been found.

Phoebe's organic surface composition is unlike any other surface yet observed in the inner Solar System. Its complex makeup differs from the rest of the Saturnian system. Thanks to the spectrometer on the Cassini spacecraft, investigators see frozen carbon dioxide ("dry ice") on the surface. An upturn in Phoebe's spectrum toward the UV may indicate scattering from iron particles. Some scientists contend that the surfaces are covered with organic molecules – polycyclic aromatic hydrocarbons. These hydrocarbons contain hydrogen, oxygen, nitrogen and carbon, the building blocks of life. The scientists assert that these volatiles might provide some evidence for Phoebe migrating in from outer regions such as the Kuiper Belt. Saturn's moons constitute a variety of bodies that probably don't come from the same place. Inner satellites formed from the rings of Saturn. Hyperion may be a fragment of another inner body, or a captured object. Iapetus may be original to the system, but its density is not what the models say it should be. Phoebe and other outer satellites are probably captured objects.

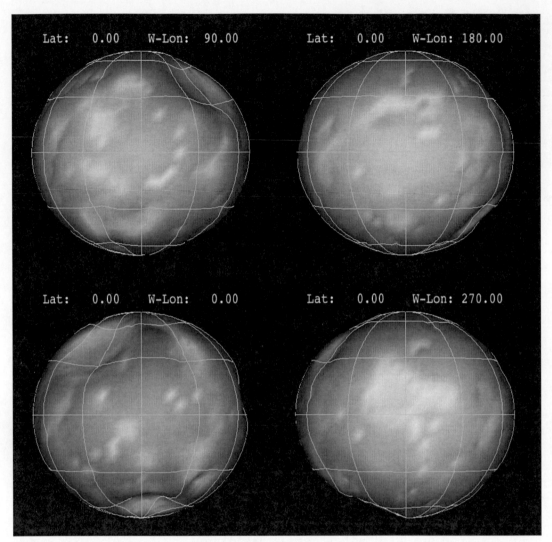

Fig. 6.15 *Cassini spacecraft images revealed Phoebe's shape to be surprisingly round for a satellite of its size (Image courtesy of NASA/JPL/Space Science Institute)*

With its cone-shaped craters, rich mineralogy, spherical shape, high density and perhaps great age, Phoebe seems more planet than moon, adding to the mystery and variety of the satellites of Saturn. But one icy moon is more dramatic than all other mid-sized satellites, playing host to towering geysers and abyssal canyons.

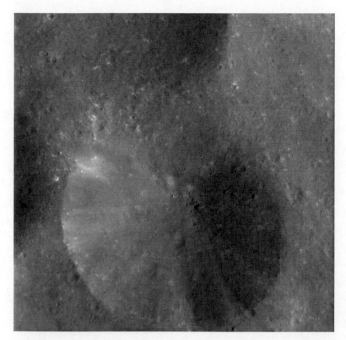

Fig. 6.16 Explorers on Phoebe will need to negotiate its steep, cone-shaped craters. This one is 13 km across (Image courtesy of NASA/JPL/ Space Science Institute)

8. That is, all craters up to 200 million years ago have been obliterated by geological forces.

ELEGANT ENCELADUS

Nestled close in to Saturn, orbiting second out of all the mid-sized moons, Enceladus circles Saturn once each 32.9 hours. The tiny ice ball has a big story to tell.

Discovered in the 1960s, Saturn's faint E-ring appeared to envelop the orbits of several of Saturn's moons, well beyond the main rings that made Saturn famous. Scientists knew for some time that a mysterious source was continually replenishing the tenuous ring. Observers also discovered that the ring was dominated by very small ice particles. Such tiny particles should only last for decades to centuries, so something had to be feeding them into the faint torus. In 1980, improved images of the E-ring revealed that its brightness peaked at the orbit of Enceladus, near the center of the amorphous ring. Some scientists suspected that the ring's fine particles were somehow related to the moon itself. They were right, but no one suspected how dramatic the relationship was.

Enceladus is an extraordinary world. Its powdery ice surface is, in places, a tortured jumble of twisted ridges and cracked plains nearly devoid of craters. These plains appear to have been resurfaced, with some areas having a geological age of less than 200 million years.[8] Other parts of the surface are heavily cratered. All surfaces of Enceladus – from ancient cratered terrain to new ridged regions – are very bright, suggesting that the entire moon is dusted with fresh material. A scant 504 km across, its diameter would barely span the country of France.

Because of its diminutive size, Enceladus' geologically young surface was mystifying. After all, other nearby Saturnian moons wore cold, dead faces, despite the fact that many are larger. The orbit of Enceladus is out of round, comparable to that of Io, so researchers assumed that tidal forces might be strong enough to trigger some kind of internal activity that affected the surface. The problem was that the moon next door, Mimas, has a similarly irregular orbit, and bears a geologically quiescent, ancient facade. Some planetary geologists suggested that the interior of Enceladus might be heated by a wobbling motion of the satellite caused by the tug of nearby moons, but many experts found this explanation unsatisfying.

In 2004, the Cassini spacecraft settled into orbit around the ringed giant. Immediately, the robot sensed the effects of the E-ring. The environs of Saturn are inundated with atomic oxygen. As Cassini coasted through Saturn's magnetosphere, it detected changes in the magnetic field lines. These changes revealed that ions from Enceladus were shifting the structure and shape of Saturn's magnetic fields, reminding some of Io's interaction with Jupiter. But the source of the disturbances remained baffling.

Positive identification of cryovolcanic activity on Enceladus came from a trio of flybys spanning February to July of 2005. The imaging science team first spotted fountains of fine mist draping curtains of light against the dark sky. Magnetometer readings confirmed the discovery, detecting ions streaming from the moon's rarified atmosphere. Enceladus' ion stream emanated from somewhere in the southern hemisphere, just where the geysers are. Detailed images revealed what appeared to be surface flows. In some locations, low-lying hills diverted the flows. The ice movement appeared similar to glaciers, although some researchers suggested the frozen courses might be thick cryolavas.

Fig. 6.17 Ice flows on Enceladus. In the image on the left, a cone-like structure, suggestive of cryovolcanic forces (arrowed), fends off glacier-like ice flows around it (Image courtesy of NASA/JPL/University of Arizona) At right, viscous deformation has wiped out some craters and modified others (Image courtesy of NASA/JPL/Space Science Institute)

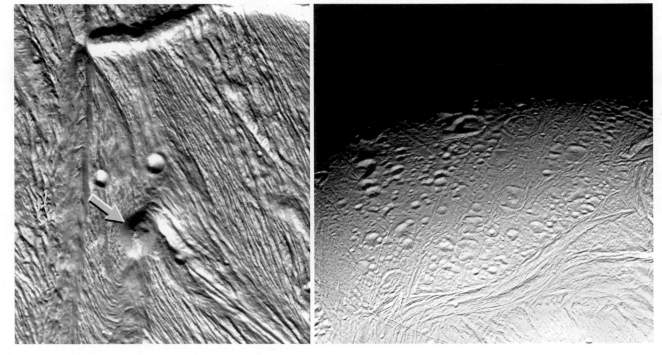

The melting point of water – and in this case, Enceladus' cryolavas – may be lowered by ammonia, which would serve as a sort of antifreeze. Temperatures in the plumes were measured as high as −57 °C (some 93° higher than the surrounding environment). This temperature is consistent with a mix of water and ammonia, which has long been one proposed mix for cryolavas at Enceladus. In several flybys, Cassini's Ion and Neutral Mass Spectrometer detected ammonia in the plumes.

Flight engineers modified Cassini's orbit. A new course carried the craft within 168 km of the surface on July 14, 2005. Team members wanted detailed data on the magnetic fields and a shot at more detailed geyser images, but they got more than they bargained for. Cassini sailed directly through an extended plume of material. The spacecraft detected 90 % water vapor, with traces of carbon dioxide, methane, acetylene, propane and possibly carbon monoxide, molecular nitrogen and whiffs of quite intricate carbon-rich molecules. Something complicated is going on in the chemistry beneath that ice. Cassini's repeated visitations give us a window into the moon's interior that we don't have for any other ice world.

In addition to all that chemical excitement, the ice particles in the plumes contain sodium chloride (ordinary table salt) and other salts. Salty ice is difficult to make unless it is flash-frozen from salt water. The plumes appear to be bringing up salty ice grains from the interior. The frozen spray may come from remnants of a long-dead ocean, but it is more likely that salt water exists not far below the surface of Enceladus right now, occasionally rocketing into the airless sky of the glittering white moon.

The plumes erupt from a series of canyons and ridges bordering a flat region in the southern hemisphere. The bizarre terrain, extending across an area at roughly 55° south latitude, consists of a snowy flat surface etched by parallel rifts. The sunken plain encircles a quartet of darkened valleys called "tiger stripes," 100-m-high ridges bracket rifts that drop precipitously some 500 m deep. Each is about 2 km across, and up to 130 km in length. Dark material extends several kilometers to each side and is apparently erupting or seeping from the rifts. The tiger stripes are roughly 35 km apart.

The surface textures of Enceladus also provide insight into its volcanic processes. Researchers chart an effect called thermal inertia, a surface's resistance to change in temperature. Cassini demonstrated that the thermal inertia of the surface in southern regions is 100 times smaller than that of solid water-ice, hinting that the landscape is "fluffy," covered in fresh ice or snow. This low-density material blankets the landscapes along the southern valleys. But daylight surface temperatures in the heavily cratered northern hemisphere are consistent with the simple action of sunlight on a solid ice surface, reaching highs of −201 ° C. Tiger stripe formations play host to much higher temperatures, soaring up to the freezing point of water. The heat is concentrated linearly along the tiger stripes. Temperatures are consistent with a heat source roughly 660 m across, which fits well with the highest resolution images of the tiger stripe canyons.

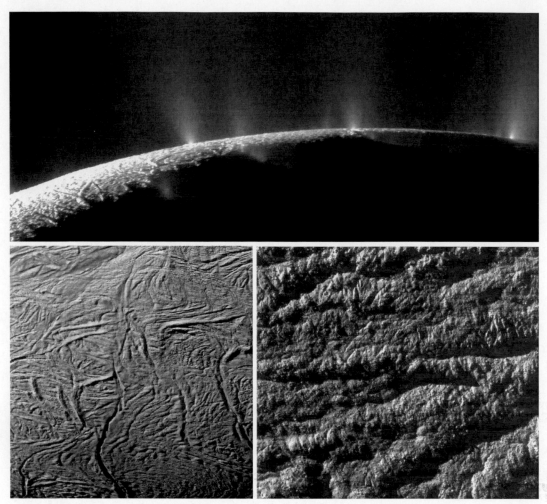

Fig. 6.18 *Progressively closer views of the "tiger stripes," source of the geysers on Enceladus. In the closest view at lower right, objects the size of a house are visible. (Images courtesy of NASA/JPL/Space Science Institute)*

The interior chemistry of the chasms fascinates geologists and astrobiologists alike. The majority of Enceladus' face is composed of almost pure water-ice. But in those tiger stripes, as in the plumes, Cassini's instruments detected organics and CO_2. Carolyn Porco, leader of the Cassini imaging team, says this makes Enceladus the best target in the Solar System in the search for prebiotic conditions or even active life. "With Enceladus, whatever it has in its subsurface ocean is there for the asking. We strongly believe the solids are flash-frozen droplets of salty liquid water that have organics in them, and who knows? They may even have microbes in them. Compounds with C-H stretch transitions have been identified in the VIMS spectra. Organics are surely along the tiger stripe fractures."

Researchers estimate that geysers on Enceladus dump 150 kg of water into space each second. A 6.5 year study yielded a total of 101 active geysers. While the material may not escape at a steady pace, the amount of water in Saturn's environment indicates the current level of activity has lasted for at least 15 years.

There is a problem with the Enceladus story. The tiny moon receives far less energy from tidal heating than can explain its energetic behavior.

Tidal heating contributes to the equation, but something else may be adding to the power – solid state friction. As Enceladus travels around Saturn, its crust flexes with the gravitational tug-o-war of Saturn and nearby moons. Fractures in the ice open and close, alternately triggering and shutting off geyser activity. The scraping of ice faces against each other also generates heat, and may enable subsurface water to make its way upward more easily.

Still, the main power behind the moon's dramatic eruptions must be tidal heating. The current best estimate of the internal tidal power is about 16 gigawatts, eight times the hydroelectric power output of the Hoover Dam. This fact challenges researchers with another conundrum, that the power output of the little moon is simply too high. Computer models based on the orbits of Saturn's satellites indicate that Enceladus can only muster one-tenth the observed tidal power on average. Enceladus may travel in and out of resonances with other moons, so that its output is currently higher than average. The moon may have been storing heat for some time and is now enduring one of its recurring active periods.

Another Enceladus enigma is why the geysers are centered on the southern pole. One possible explanation is that the heating started in another location, and that material was transported away from the heated area either by melting of the ice or by loss through eruptions. When a spinning object loses mass in one area, its rotation becomes unstable until the spin axis realigns itself into balance. The active region on Enceladus may have begun somewhere else and then realigned itself with the pole, becoming stable again. An ice crust that is floating on a sea of liquid water can move much more easily, and generate more frictional heat, than ice that is frozen solid to the rocky core beneath it. Enceladus likely has a subsurface sea now, or did in the recent past. With a subsurface sea, complex organics and low radiation compared to the Galilean satellites, Enceladus will be a prime target in the search for extraterrestrial life.

With the advent of sophisticated missions such as Galileo and Cassini, planetary scientists are beginning to understand the icy satellites as a cohesive group, an interrelated family of associated bodies. The barrage of revelations beamed back from the Voyagers was nearly overwhelming, but began this process, says John Spencer. "The idea that the moons of the outer planets were active in recent geologic time was a huge revelation. With Cassini discovering the activity on Enceladus, it really changed so much of what we knew there, but we were already open to that possibility, because of what Voyager had shown us and the fact that we'd seen activity on Io, and we had hints that there might be activity from the Voyager data and other data from Hubble and so on. So it was amazing and fabulous and made sense of so many things that were quite puzzling up to that point, but it wasn't completely out of left field the way that finding volcanoes on Io was. We knew that such things were possible."

Voyager brought a sense of pattern to the variety of small icy worlds, but within those patterns and trends, the puzzles continued to challenge observers. "I was surprised simply because parts of Enceladus

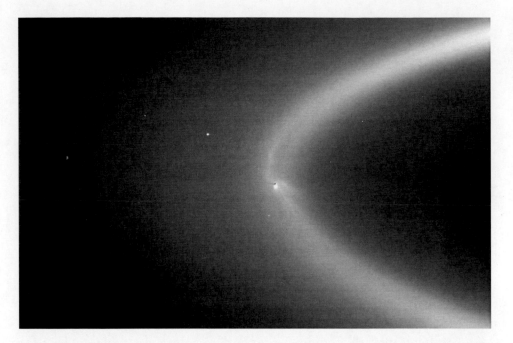

Fig. 6.19 Cryovolcanism on Enceladus creates Saturn's amorphous E-ring (Image courtesy of NASA/JPL/Space Science Institute)

that you could see from Voyager were pretty old. Io has no craters on it. Europa has very few craters. But Enceladus was different. Here you could have a world that combined these very old regions where not a lot had happened for a billion years with these very new regions where things were happening last week."

Researchers including William McKinnon see hints of that same activity elsewhere. "Enceladus has been active in other regions in the past. If you really push it, it looks like Dione has been active. If we could just figure out how Enceladus works, what the plumbing is like, how extensive the water is down there, how it all works together, we could have a deeper understanding of the other satellites. For example, if you take Enceladus' tiger stripes to a higher gravity environment, maybe the result is something like Ganymede's grooved terrain."

Scientists continue to probe the mysteries of the icy worlds, assembling theories and searching for trends in the bizarre histories, origins and evolutions of these important moons. Research will continue, both robotically and, one day, with human explorers carrying the future equivalent of a geologist's hammer. One thing will not change, as Spencer observes: "We continue to be surprised."

Tiger Stripe Tour with Carolyn Porco

Cassini Imaging Science Team leader Carolyn Porco has scientifically informed dreams of a day when travelers might venture to the snows of Enceladus. Here are her thoughts:

> I got interested in the icy satellites by being on the Voyager imaging team and looking at all the fabulous results, thinking of them as worlds, but I really got interested in studying them scientifically from my team's discovery of the geysers [at

Enceladus]. I did rings for thirty years. I did my thesis on Saturn's rings, and then came work on Uranian rings and then Neptune. Before the Cassini mission, I was convinced that the excitement was going to be at Enceladus. Back in our team proposal, I said "Enceladus is going to be the Europa of the Saturnian system." But we never expected it to be so dramatic.

I've said over and over that Enceladus is the most accessible habitable zone. It would win hands down over Mars or Europa. Europa is bathed with such intense radiation that it's hard to imagine biology. If organic materials are found on the surface, how do you know that it's not organic material raining down on the surface? But with Enceladus, what's inside is accessible. You don't have to scratch. You don't have to dig. You don't have to sniff. You don't have to do any of those complex things. All you really have to do is land on the surface, look up and stick your tongue out. Roughly 90 % of the solids go up and come down again.

That's one problem with sending humans to Enceladus: you don't want to contaminate it. Still, I long for the day when distant descendants visit the Enceladus Interplanetary Geyser Park. Think about standing in a forest of jets that tower hundreds of kilometers above you. And with Saturn and its rings? I'm sorry; Jupiter is great, it's colorful, but Saturn and its rings? [On the Galilean satellites], you would be wearing lead goggles. You wouldn't see much!

Getting into orbit takes some doing, because Enceladus is so light – not dense – but you can bleed off lots of energy with lots of flybys of Titan and Rhea. Once you get there, it's trivial to land on. It's almost like landing on a comet. In fact, if you made a mistake, you'd have time to correct yourself because things happen so slowly. You want to be near the tiger stripes. You want to be down in one. You'd have to have Neil Armstrong built into your spacecraft, but it's doable.

Being on the surface has to be carefully controlled so you don't contaminate the environment. It might make sense to have an orbiting depot. You wouldn't have to worry about fuel to get in and out of the gravity well. You could build it in orbit around Earth and then transport it out there.

FUTURE TRAVELS

What sort of landscapes will future astronauts see on these small ice worlds? John Spencer surmises that the surfaces of these moons will be reminiscent of Apollo landing sites. "I suspect Tethys, Dione and Rhea will look quite lunar from close-up, the ice being fairly homogeneous and behaving like rock (unlike on the Galilean satellites, where sublimation is important) and the surfaces dominated by craters on a small scale, except for the occasional giant cliff. On the leading side of Tethys the surface will be particularly 'crunchy' due to electron bombardment, and on the trailing sides of all three, especially Dione, there's a layer of dirtier ice that covers everything except the ice cliffs."

Visual similarities between icy satellites and Earth's Moon come at the hands of similar cosmic erosion. The primary cause of erosion on Earth's moon today is from a constant drizzle of micrometeorites, which wear down the surfaces over long periods. Harrison Schmitt, geologist and crewmember of *Apollo 17,* underlined the ubiquitous nature of micrometeorites in his first-hand observations. He said, "When the Sun is behind the observer, the regolith sparkles like snow. Boulders of every size and shape are strewn everywhere. And all of them, down to the small rocks at your feet, are covered in tiny white dots from micrometeorites. The landscape is pocked with craters of every size, down to small rimless bowls, each with a fused-glass center."

JPL's Bonnie Buratti concurs that micrometeorites will be the primary sculptor of outer planet landscapes. "Most of what we have learned so far is that collisional processes in the Solar System seem to work the same on icy and rocky bodies."

The icy satellites may look strikingly different from our own Moon at a distance, but planetary geologist Geoff Collins warns that appearances may fool us. "Even in places that look, from a distance, like they might be fresh cliffs and ridges, once you get up close they look like they've been softened by a continuous rain of impactors. It's like we see on the lunar surface. There were some craters that looked fairly fresh before the astronauts got up there, and then once they were up there it's all covered by this dusty mantle. There were sharp changes in slope that made things look fresh from a distance, but it was all still softened."

With its dramatic geysers and proximity to Saturn, some futurists consider Enceladus prime real estate for future travelers. But NASA Ames' Jeff Moore suggests that Iapetus is also well placed for human exploration: "Ideally you should be in the outer part of the system – for example, Iapetus – because the other moons are always relatively near the primary. You always have this great view of this gas giant and its moons from your convenient outer-moon vantage point, maybe on top of that equatorial ridge. You can imagine that after the millionaires have paid for their trip out to Saturn in the first place, they can board some sort of one- or two-day flyby and go zoom around the canyons of Enceladus and come back."

Planetary geophysicist William McKinnon agrees, adding, "I could see climbing the ridge of Iapetus for a great view. It's a 20-mile hike, but the gravity is low."

Another advantage of an Iapetus outpost is the moon's relationship to Saturn's spin axis. Nearly all of Saturn's moons orbit very close to the ring plane. Frustratingly, this vantage point presents a view of the glorious rings edge-on, and those rings are quite thin. From most of its satellites, Saturn's spectacular ring system would be compressed into a razor-sharp line against the sky and planet. Not so Iapetus. Its orbit is inclined 15½° to the ring plane. During the course of its 14-day circuit around the planet, the moon's path carries it first above and then below Saturn's rings; the planet appears to oscillate like a slowly spinning top. In some places, Iapetus' towering ridge would provide a slope – facing always toward Saturn – on which to perch a scientific outpost and, eventually, a vacation resort for that well-earned distant getaway.

Fig. 7.1 A future visitor watches the surf on the shore of Kraken Mare, a methane sea of Saturn's moon Titan (Painting © Michael Carroll)

Chapter 7

Titan, the Other Mars

Earth's Moon	Europa	Mars	Earth	Titan	Venus
~0.0 bar (vacuum)	0.000000000001 bar	0.006 bar	1.0 bar	1.5 bars	89 bars

atmospheres compared

Fig. 7.2 Titan's chemically rich atmosphere is the second densest of any solid body in the Solar System (art by the author)

If one grabbed the planets and their moons and scattered them across a starry table, arranging them by size, Titan would fall somewhere between the planets Mercury and Mars. It has more atmosphere than either planet. In fact, Titan has more air pressure than Earth does.

Titan is unique among moons in our Solar System, too, as the only moon to have more than just a trace atmosphere. Its opaque nitrogen-methane cocoon is the second thickest among all the solid bodies of the Solar System. At 1.5 bar, its atmospheric blanket sustains a surface temperature of −178 °C, much warmer than nearby Enceladus, whose daytime temperatures hover around −201 °C.

Titan's year – identical to Saturn's – lasts a long 29.7 Earth years. Its environment is nearly as complex as Earth's, with dynamic meteorology and an active hydrological cycle unlike anything else found throughout the Solar System, with the sole exception of our own world. In fact, says planetary scientist and engineer Ralph Lorenz, "The principal reason that we're not more interested in Titan than we are is that it's so far away. It takes 7 years for a spacecraft to get there. Were that obstacle removed, either by moving Titan closer to the Sun or with better propulsion technology, I think it would be a target of great interest for a number of reasons. First, it's a very rich environment phenomenologically." He continues: "Because it has a thick atmosphere, there are processes going on there that we don't see on Mars today – the formation of clouds and rain, the pooling of liquids on the surface, their movement by tides and presumably by wind as well, the formation of waves. You have a much more astrobiologically interesting chemistry at Titan because of the

M. Carroll, *Living Among Giants: Exploring and Settling the Outer Solar System*,
DOI 10.1007/978-3-319-10674-8_7, © Springer International Publishing Switzerland 2015

Fig. 7.3 From an altitude of ~18 km, ESA's Huygens probe could make out branching river valleys in the icy landscape below (Image courtesy of NASA/ESA/JPL/University of Arizona; images processed by Rene Pascal)

methane in the atmosphere being processed into literally hundreds of other compounds."

Researchers suspected that Titan held great promise in these areas early on. Biophysicist Benton Clark yearned to be part of Titan exploration even back in the 1970s, when he was designing instruments for the first successful Viking landings on Mars. Back then, Titan was an unknown, and designs for Titan landers took into account a range of possibilities, from snow banks to solid surfaces to methane baths. Decades later, Clark's yearnings became reality when he took part in the European Huygens Titan probe project. Huygens piggybacked aboard NASA's Cassini Saturn Orbiter, arriving at Saturn in 2004 (see Chap. 3).

"This was especially exciting because you were going to such a different place," says Clark. "It was thought, originally, that Titan would be covered in an ocean. Subsequent data showed that it was mainly land. But shortly after we did land, we'd learned that there are large seas found on Titan. [Huygens] was exciting because it was a whole different mission compared to a Mars landing, and it was European. The Europeans designed it. They had never landed anywhere before, so there were all the elements of excitement, chance, the first time doing something."

After a 2½ h parachute descent, Huygens set down on terra nova, a new world in every sense of the term. The probe radioed back a precious treasure-trove of in situ data on Titan's complex layers of air, clouds and mists, and gave us aerial and surface views of the alien landscape below. The plucky probe continued to relay data for 90 min after touchdown before contact was lost.

Beneath its obscuring orange smog, Titan's frontiers continue to be laid bare by the Cassini orbiter's prying instruments.

Fig. 7.4 Huygens continued to transmit images of this swath of Titan's surface, an apparent dry riverbed, for 90 min. Note the rounded stones, typical of eroded river rock (Image courtesy of NASA/ESA/JPL/University of Arizona; images processed by the author)

Aside from some baffling terrain, many regions in those first radar passes looked hauntingly familiar, Clark says. "It looked like geology on Earth to a surprising degree. You had these flow features. There were river valley systems that appeared to be draining into things that looked like they could be lakes." Additionally, Titan plays host to mountain ranges and great sand dune seas, fine itineraries for future explorers.

TITAN'S ATMOSPHERE

Obscuring those canyons, lakes and dunes, Titan's murky blanket of nitrogen and methane has its own story to tell. On both Earth and Titan, weather is simply nature's way of trying to balance out temperature and pressure. Heat comes in from the Sun, and wind carries the warm air to colder areas. Just 149 million km from our own world, the Sun pumps prodigious amounts of energy into Earth's system constantly. This makes our atmosphere a vigorous and active place.

The solid surface beneath our blanket of air breaks the flow and disrupts temperature gradients, so that the environment cannot maintain weather systems for extended periods. Our capricious meteorology stands in stark contrast to that of distant, frigid Titan. Our numerous thunderstorms, thrust forward by cold fronts and held at bay by warm fronts, scatter the white banners of clouds so familiar to fans of the Weather Channel. Earth's rainstorms take warm air near the surface and loft it to higher altitudes.

On Titan, it doesn't work that way. Titan receives one-hundredth the amount of sunlight that Earth does. Less sunlight enters the atmosphere, causing more gentle mixing of the air, which means that there are fewer individual weather events. An occasional storm may be all that occurs over the course of many years.

Analysts have likened atmospheric heating to a pot of water on a stove. When the flame is first lit, an occasional blob of water will rise from the bottom of the pan to the top, causing little turbulence. As time goes on and the water gets hotter, more and more of those blobs will rise through the liquid. The water in the pot can be very turbulent even before the water itself begins to boil. Titan's atmosphere may be similar to the beginning stage, where blobs of warm fluid come up only rarely. Earth's atmosphere is similar to the pot's liquid at the boiling stage.

The rarity of Titan's dramatic weather, coupled that with the fact that Cassini observes Titan infrequently for short periods, makes for challenges in discovering significant weather events. Still, Cassini has spotted evidence of convective clouds in the south. In particular, lake patterns in the southern region near Ontario Lacus have changed in several places. Observers suspect that methane storms may have produced new lakes in that area.

Fig. 7.5 Titan's complex haze layers lead to methane rain and hydrocarbon precipitation that results in dunes (Image courtesy of NASA/JPL/SSI)

Titan's rains may be monsoonal; rainfall may come in intense, seasonal waves. Many river channels have been charted across Titan's surface, and these typically need intense floods to produce them. Planetary meteorologists reason that a steady drizzle won't do it. Titan's methane storms may be most like storms over terrestrial deserts that wet the ground and even carve arroyos but don't result in any significant buildup of bodies of liquid.

Despite active methane cloud systems and carved floodplains, it appears that methane rain falls far less often on Titan than water rain falls on Earth. Especially in the equatorial regions, Titan is very much a desert planet, where rainfall is extremely rare compared to Earth. Where there is precipitation, it is torrential. The question is, is that torrential rainfall seasonal, or is it a year-round phenomenon? Although the methane humidity at the equatorial Huygens site was 45 %, which would be enough to trigger rainstorms on Earth, the distant Sun's heat at Titan is just too weak to drive moist air up and generate storms under current conditions. Large cloud systems would require higher humidity. Titan's equatorial region is just too dry.

But not completely. In 2008, Cassini spied storm clouds forming over the equatorial regions. They blossomed rapidly to the southeast. As the storm clouds drifted away, they left in their wake a darkening of the landscape, implying a changed surface from rainfall. This phenomenon may be rare, but Titan showed us that it does, nevertheless, happen.

Researchers estimate that the annual rainfall on Titan amounts to about 5 cm. This is the equivalent to annual precipitation in Death Valley. That rain falls upon a host of truly bizarre landscapes.

Titan's atmosphere provides a window into deep time. Most scientists assumed that its nitrogen formed within the warm protoplanetary disk that formed Saturn itself. But new research by NASA and ESA points

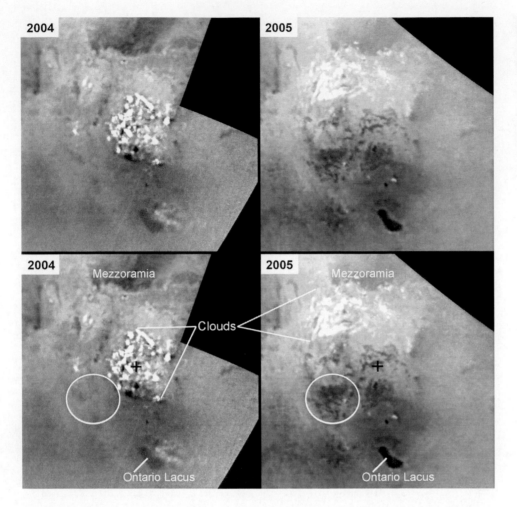

Fig. 7.6 White methane storm clouds drift over gray surface features in these Cassini radar images taken in the region of the great southern lake Ontario Lacus. Note changing patterns within the dark areas on the surface as the bright clouds glide overhead (Image courtesy of NASA/JPL/SSI)

to a more distant source – the outer edge of our Solar System. Titan's nitrogen has a ratio of two isotopes, nitrogen-14 and nitrogen-15, that more closely matches isotopes that should originate in the cometary ices of the Oort Cloud, which forms the outer fringes of the Sun's influence. This means that Titan's nitrogen formed very early in the Solar System, predating the formation of the planet it now orbits. The fingerprints of conditions in the early Solar System may be preserved in Titan's atmosphere today, making it an important target for scientists studying the evolution of planetary systems.

ALIEN DUNES

From the start, Cassini's VIMS (Visual and Infrared Mapping Spectrometer) and ISS (Imaging Science Subsystem) instruments revealed dark regions across the face of the veiled moon. More detailed radar sweeps resolved linear features covering at least some of those dark areas. The Applied Physics Laboratory's Ralph Lorenz says that those first images were misleading.

"In some ways we were quite unlucky with the bits of Titan we saw first. We were putting labels on things like 'cat scratches.' The very first radar flyby, in Oct of 2004, was pretty inscrutable."

However, on Cassini's second radar flyby, acquired just a few weeks after the spectacular Huygens descent, more interpretable structures came to light. The radar path included two large impact craters and a distinct set of branching river channels. "The Huygens probe had already seen river channels at a much smaller scale," Lorenz remembers, "so that discovery, if you like, had already been scooped by the Huygens probe by historical accident. But also on this flyby there were all these dark streaks, and they're not perfectly parallel in that area. They often branch in successive "Y" junctions. We suspected they might be some sort of Aeolian – wind-blown – deposit, but we couldn't be sure they weren't some other kind of surface flow like a river deposit. They were obviously different from the things that looked like rivers, but we couldn't be absolutely sure."

One of the theories researchers were working on had to do with some radar structures seen in Antarctica called megadunes. These are not mountains; they have subdued topography to almost none at all. But to a large extent, Lorenz and his colleagues were distracted by the things that they did recognize, so they concentrated on the craters and the river channels. It was only another flyby about 5 months later (October 2005) that they looked at a part of Titan closer to the equator than the earlier passes, where previous telescopic observations revealed large, dark patches. The new radar image was almost completely covered with these linear features, and now they weren't just dark streaks on a brighter background, Lorenz says. "We could see from the pattern of bright and dark in the radar image that there were slopes facing toward the radar and slopes facing away, so we knew these things were ridges. They were very uniform, a couple of kilometers apart, twenty to hundreds of kilometers long in many cases." He added, "Because they were elevated, that first gave us a clue that they were depositional dunes, and in fact they turned out to be. The moment it finally clicked was looking at a space shuttle picture of the Namib sand sea where, even though the sand is different and gravity is different and the air is different and all that, you see exactly the same thing. These large linear dunes 2 km apart, 100 or 200 m high, tens to hundreds of kilometers long."

During Cassini's first 4 years at Saturn, scientists estimated that dunes might cover up to 10 % of the surface. With more scrutiny, that has changed, according to planetary sand dune expert Jani Radebaugh of Brigham Young University. "We now estimate that at

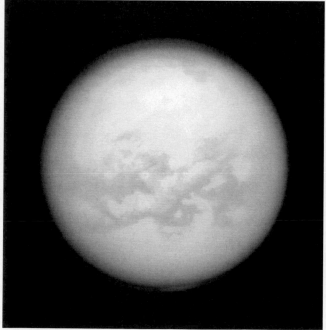

Fig. 7.7 Great sand seas spread across Titan's equatorial regions, appearing to Cassini's imaging system as dark swaths in the fog (Image courtesy of NASA/JPL/SSI)

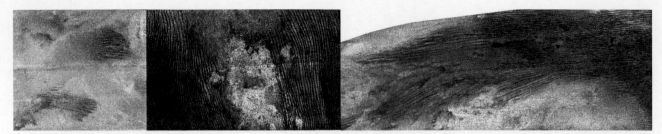

Fig. 7.8 Cassini's "cat scratches" (left) resolved into linear dunes as more radar coverage came in (Images courtesy of NASA/JPL)

least 20 % of Titan's surface is covered by dunes, and perhaps much more." By comparison, dunes cover about 5 % of Earth's land, and less than 1 % of the Martian landscape. Titan's long dunes act as a sort of weather vane, following the direction of the moon's winds. Titan's atmosphere is 1½ times as dense as Earth's at sea level. Chilled to 143° below zero C, the sluggish air moves across the face of Titan like a planetary tidal wave. The shape and orientation of the dunes indicates that Titan's winds blow from west to east. By February of 2009, researchers had mapped 16,000 dune fields to confirm the findings. More would come.

Titan's dunes may be one of the most alien features of the planet. To understand their nature, one should look only as far as Titan's surface makeup, says Ben Clark. "The surface of Titan has rocks, but the rocks are thought to be water-ice. You have water available, but it's frozen 'rock solid.'" If they were analogous to Earth's dunes – silica sand that comes from ground up rock – Titan's dunes should be ice. But Titan's dunes may not be comprised of water-ice at all. They may, in fact, be composed of organic material that falls from the sky. Cassini's Visual and Infrared Mapping Spectrometer sees all the dunes as dark. If they were water-ice, they would appear bright. Cassini's radar offers another clue. In addition to providing a visual image, radar yields insights into how the material behaves. Radar waves bounced off the surface measure a dialectric constant, data that tells scientists about the size and makeup of the material. Titan's dunes yield a dialectric constant that is not consistent with water, but instead indicates fine-grained organic material. This soot-like hydrocarbon matter precipitates out of the sky as a result of the interaction of the Sun's ultraviolet radiation and methane in Titan's atmosphere.

"The material is abundant," says Lorenz, "but it's had quite possibly a billion years to accumulate. The amount that falls down in any one day is tiny. We don't think the dunes are made of exactly the same stuff that makes Titan's atmosphere orange and opaque. The photochemical material that forms in the atmosphere appears to drizzle down as a smog of very fine particles, a third of a micron or so, so it's very similar in some ways to smog. But it's not really falling down in the sense of a snowstorm. You're not seeing it accumulate. Now, the stuff that makes the dunes appears to be darker in color, and it must be in larger particles."

One idea recently put forth is that the hydrocarbons drift out of the sky and land in the seas. The density of liquid methane and ethane is such that almost anything sinks in it. "You could imagine something very fluffy

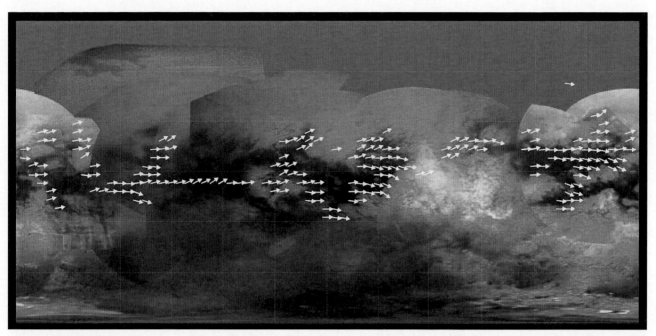

Fig. 7.9 Titan's dunes act as global weather vanes. Arrows indicate wind direction on a planetary scale (Image courtesy of NASA/JPL/Space Science Institute)

like pumice that could float," says Lorenz, "but any imaginable bulk material would sink in Titan's seas. Things would sink very slowly, because the gravity is low and the density difference between the particles and the liquid is low, so think Miso soup."

The particles eventually settle out, but even convection currents are enough to lift the particles up. This means there are a lot of opportunities for particles to clump together in Titan's seas before they settle. Over long, glacial-like timescales, Titan's seas basically dry out. These long cycles are called Croll-Milankovich cycles. The cycles depend on the leisurely shift of a planet's orbital trek around the Sun.

In Titan's case, its Croll-Milankovich cycles are tied to those of Saturn itself. As Saturn's orbit precesses, the seasons on both Saturn and Titan lengthen in one hemisphere and shorten in the other. This change might lead to a drying out of seas in one Titan hemisphere, and a shift in methane humidity to the other. We find a similar phenomenon recorded in Earth's geological record. Huge layers of salt lie beneath the Arabian and Mediterranean gulfs, because they were once ocean basins that became closed off, eventually drying out.[1]

In this process of drying out, fine hydrocarbon particles in a Titan sea may come together as grains to make sand. The sand gets blown out and makes its way to the equator. Titan experts see a global sediment "budget" in which sand is created, migrates and eventually disappears. This planetary sediment picture remains one of the big mysteries of Titan. In fact, it is very much a Mars-like question.

Thanks to long-term studies by spacecraft and to our computer models, we've gained insights into Mars' climatological past, influenced by the same kind of Croll-Milankovich cycles that Titan is subjected to. The Croll-Milankovich periods on Mars are estimated to cycle every

---

1. The Mediterranean and Arabian gulfs are, of course, filled in with water today, and this may make them an even better analogy for some lakes in Titan's north.

50,000–100,000 years. The same changes in the eccentricity and obliquity of the orbit, and changes in how much sunlight is available in the summer, are expressed in the dust and ice layers of the Martian polar caps. And if Titan's rivers are dormant today, those channels are analogous to Martian river channels that could not be made under current conditions. Many Mars experts are convinced they are evidence of a different climate on Mars, a warmer, wetter past. Now, investigators are beginning to have feelings of *deja vu* about Titan, even though Titan's present climate is far more dynamic than Mars' present climate, with its active hydrological cycle. Thanks to recent data, researchers are able to start thinking about Titan's evolution through time.

With a constant global drizzle of organic particles, Titan should be covered in dunes from north to south, but this is not the case. In fact, most of the dunes appear in a belt within about 25° or 30° of the equator. Their location appears to be primarily a function of Titan's atmospheric circulation. Most of Earth's dunes exist in *two* belts about 20° or 30° *from* the equator, because Earth rotates fairly quickly, and the warm rising air over the equator heads off toward the pole but gets sheared off by the rotation. Lorenz

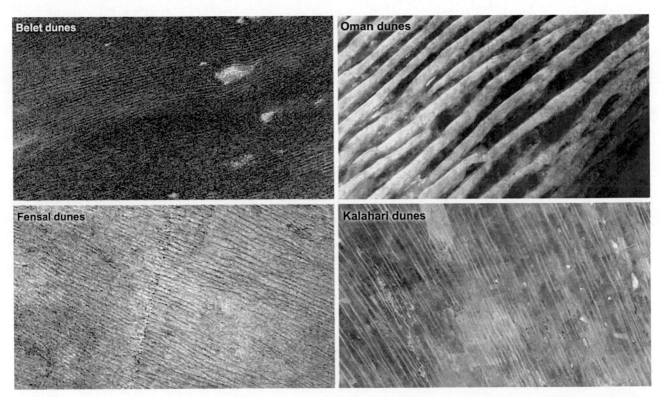

Fig. 7.10 At higher elevations or higher latitudes (bottom left, Fensal), Titan's dunes tend to be thinner and more widely separated, with gaps that have a thinner covering of sand. Dunes in the Belet region (top left) are wider, with thicker blankets of sand between them. The Kalahari dunes in South Africa and Namibia, located in a region with limited sediment available (bottom right), show effects similar to the Fensal dunes. The Belet dunes on Titan resemble Earth's Oman dunes in Yemen and Saudi Arabia, where there is abundant sediment available (top right). The altitude effect suggests that the "sand" (likely composed of hydrocarbons) needed to build the dunes is mostly in the lowlands of Titan. The image of the Oman dunes, also known as dunes in the Rub' al Khali or Empty Quarter, was obtained by the Advanced Spaceborne Thermal Emission and Reflection Radiometer (ASTER), an instrument aboard NASA's Terra satellite (Images courtesy of NASA/JPL-Caltech; NASA/GSFC/METI/ERSDAC/JAROS; and U.S./Japan ASTER Science Team)

explains: "The down-welling branch of air is at about 20°–30°, and because it's had clouds form in it, the air that's coming down is dry. That's what leads to the accumulation of our deserts here on Earth. On Titan, because the rotation is much slower, you get the drying in the equatorial belt." This leads to Titan's massive areas of dunes. The largest dune "sea," known as Belet, stretches some 3,000 km in length. Its dunes tower 150 m high.

On Earth, coastal dunes snake along shores of lakes and oceans. No such dunes have been found on Saturn's moon, says Jani Radebaugh. "That's a really interesting thing we need to puzzle through. There are no dunes above 60° latitude." Dunes need sufficient winds and sediments to form. They also need fairly dry conditions, so Radebaugh suggests that the lack of dunes in polar regions, where evidence of methane lakes has been found, may be the result of elevated levels of methane vapor.

Radebaugh has identified a few sporadic dunes that are further from the equator. They appear to be more dense. Researchers associate radar dark substances with low density, like the organic fallout. Radar bright material tends to be icy or rocky. These outlying dunes look brighter, suggesting they have a different texture. One possibility is that these are frozen dunes, somehow cemented together. They also have a different orientation from the main sand seas in the equatorial region, so one theory that has been proposed is that they are left over from a previous climate epoch when winds tended in a different direction, making their orientation different. Perhaps the humidity was different in past epochs, enabling sands to move farther north, further fleshing out the Croll-Milankovich picture.

The source of that sand is tied to the past, and a planet like Mars or Titan may have a very long memory. Dunes may reflect history going back millennia or even millions of years. For example, a large dune may take tens of thousands of years to form under certain wind and weather conditions. But as those conditions change, the dune orientation and shape will be askew in relation to current climate. Smaller dune features form and change more quickly, and more accurately reflect today's conditions. Lorenz points out that Mars has a similar situation with its sand dunes. "Large dunes may be Mars' memory of how the winds once were. Similarly, close study of Titan is starting to hint that while the dune pattern overall is very regular, reflecting a mature 'end state,' small features at the edge of some dune fields may indicate a changing climate."

Both Lorenz and Radebaugh suspect that Titan's dune fields are active today. Radebaugh points out that Cassini's radar has not been able to detect specific changes, but she did not expect it to.

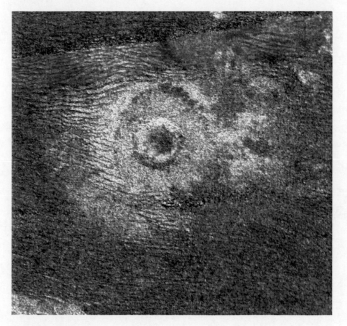

Fig. 7.11 Sand dunes encroach upon a large crater 40 km in diameter (Image courtesy of NASA/JPL/Caltech/ASI)

Our resolution is too low. Even on Earth, where winds are higher, seeing movement of 170 m over a few years would be a stretch. But there are other indicators that these dunes are some of the youngest features, and likely even active. We know there is rainfall, and yet we do not see strong evidence of river channels cutting through the dunes. Even in Namibia the big, dry river valleys carve through the dunes, which are being blown by winds now. Also, the inter-dune areas are swept clean, as seen by VIMS, yet if they were just undergoing mass wasting and redeposition, the inter-dunes should be covered, like in the Nebraska sand hills. I side with the dunes moving now.

The formation and transport of Titan's bizarre hydrocarbon dunes is dependent on the dynamics of the moon's dense air. And although the majority of Titan's atmosphere is nitrogen, it is its second most abundant constituent, methane, that's the driving force behind chemistry and weather. Because sunlight destroys methane, the gas should be short-lived in Titan's environment, but it isn't. Something must be replenishing it.

CRYOVOLCANOES AND OTHER LEAKY SOURCES

Consensus is building in the planetary science community that Titan's interior pumped methane into its skies during three developmental epochs. In its formative years, as the moon accreted from the solar nebula, a rocky core formed beneath a water mantle. A water-ice crust topped the mantle. During its first several hundred million years, heat from the moon's formation combined with the warmth of radioactive elements in the core to melt through the crust, releasing methane.

The second release probably occurred about 2 billion years ago, when Titan's silicate core began to convect. This geological burst of heat again melted the crust, causing methane outgassing. Ammonia mixed with the water-ice would have helped to serve as a natural antifreeze, perhaps enabling widespread cryovolcanism.

As Titan settled down after this violent era, a mix of methane and water-ice would have formed a lattice, called a clathrate. This clathrate crust would gradually thicken above a layer of pure ice. Convection would have begun within that outer crust itself, freeing the methane trapped within as geyser plumes or gas leaking out of the ground.

Whether the details of Titan's planetary evolution followed this path or not, many believe that methane must be escaping today, recharging the atmosphere at a fairly steady pace. But if so, where is it coming from?

Fig. 7.12 Two possible cryovolcanic sites. Left: Flows scar the landscape surrounding Hotei Arcus (Image courtesy of NASA/JPL/SSI) Center: Location of Sotra Patera, a circular mountain structure, with its computer 3-D reconstruction (far right) (Image courtesy of NASA/JPL-Caltech/ASI/USGS/University of Arizona)

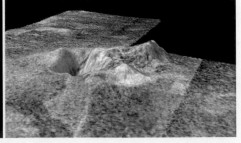

The search is on for evidence of outgassing in the form of cryovolcanic formations. Scientists at the Jet Propulsion Laboratory have been studying several features that resemble volcanoes, including an area called Hotei Arcus. "VIMS sees brightness changes in two regions," says JPL's Rosaly Lopes. "In both Hotei Regio and western Xanadu, RADAR sees flow morphology consistent with cryovolcanic flows." From October of 2005 to March of 2006, Cassini flew by Hotei Arcus three times. Lopes and others believe the VIMS images show surface darkening from one encounter to the next. One likely reason – cryovolcanism, super-cold volcanic activity.

Hotei has also been under scrutiny by Randall Kirk at the U. S. Geological Survey in Flagstaff. Kirk is skeptical about the VIMS data, but he still feels Hotei is likely a volcanic flow feature. "The first look we had in radar showed the kind of lobes and protrusions and indentations that you get with lava flows or other viscous material. Some thought these might be sedimentary deposits related to the narrow channels that were flowing into the area." Then, Kirk's team got a second radar pass, enabling them to construct stereo images of the area. "What we found is that they have a measurable thickness of 100–200 m; the channels come into the base level."

The new data suggests that the channels could not have deposited the flows, which tower above the surrounding landscape. Several other sites may display possible cryovolcanic signatures. Some features seem to have a hole in the ground, like a caldera or volcanic crater, with a thick, serpentine flow issuing from it, similar to silicate lava flows. Another region, Tui Regio, has the same unique spectra that Hotei does and seems to be composed of similar flow-like features. Like many formations on Titan, some flows are confusing and enigmatic. Clearly, something flowed across the surface, but the flow patterns are so diffuse that they could be thin or thick, and could be volcanic or caused by methane rain outwash.

SCIENCE AND CAVEATS

Titan research provides plenty of cautionary tales. Early in the Cassini mission, a remarkable structure came to light that reminded investigators of volcanoes on other worlds. Radar images uncovered a 180-km-diameter circular feature that mapmakers christened Ganesa Macula. Although no topographic data was available at that first pass, the mountainous feature bore a strong resemblance to volcanoes on Venus or Earth, specifically pancake domes or shield volcanoes.

Ganesa is about the same scale as large shield volcanoes on Earth. The dome appeared to have a relatively flat top, steep flanks and a central depression about 20 km across. Researchers interpreted this depression as a volcanic caldera. Sinuous channels wound their way down the mountainside from the central crater, looking for all the world like cryolava channels running down the flanks. Additionally, flow features that appeared to have erupted from the dome were preferentially located on the south and eastern sides, implying a gradient. One volcano expert commented,

"The morphology of putative flows emanating from the south and east sides of Ganesa is not indicative of viscous flows. They are instead thin, sheet-like and broad." If Ganesa was, indeed, the smoking gun for Titan volcanism, its flanks appeared to have been laced with "lava" flows of water, which slumped down the mountainside like terrestrial glaciers.

A second radar pass several years later, however, enabled researchers to assemble a stereo image of Ganesa. Rather than a domed profile, the "mountain" turned out to be nearly flat. Its flows and morphology remain a mystery.

Nevertheless, ESA's Huygens probe obtained other clues suggesting the presence of volcanism. Its amazing surface images did not show any features that were unambiguously cryovolcanic, but Huygens detected the radiogenic isotope of argon (40Ar) in Titan's atmosphere. Its presence

Fig. 7.13 Cassini's first radar image of Ganesa Macula, top, led observers to the conclusion that the feature might be similar to the pancake volcanoes of Venus. Below is an early – and incorrect – interpretation (Top image courtesy of NASA/JPL; bottom painting is by the author)

indicates that the atmosphere must be in contact with subsurface potassium. It is likely that silicate rocks make up most of the potassium-bearing material in Titan's core. Cryovolcanism would be one means by which this material might be brought to the surface.

MOUNTAINS

Whether the structures discussed above are volcanic or not remains to be seen. Skeptics suggest that Hotei and other theorized volcanic sites may instead be products of uplift or other mountain-building processes. But Titan does display other types of mountains, equally baffling.

To many, Titan's summits came as a surprise. The only two other moons of similar size, Callisto and Ganymede, are dominated by craters and, in Ganymede's case, grooved terrain. Scientists predicted that any mountains on Titan would be remnants of impact crater rims or central peaks. But Cassini has revealed parallel ridges, solitary raised structures, and mountain chains. Geologists have offered four possibilities for the origins of these mountains:

1. The mountains were thrust up from below as two areas of crust pushed together in a process called compression.
2. Two sections of crust separated; one remained while the other dropped, forming a graben or horst, ridges left along sunken terrain. This process, called extension, may be prevalent on Ganymede (see Chap. 5).
3. The mountains are remnants of blocks of material ejected from large impact craters.
4. Erosion stripped away material from a preexisting plain, leaving mesas. (Watch for this process on Uranus' moon Ariel in Chap. 8.)

Some or all of these processes may be at work, says Jani Radebaugh. "It depends on where you are looking. There are long chains…that are parallel and trend east/west. We're leaning toward compressional formation for those." But 200 km from the crater Sinlap lie several mountains of an entirely different nature. At first blush, these peaks appeared to be oriented radially from the crater's center, as if they were products of impact ejecta from Sinlap itself. Calculations show that in Titan's thick atmosphere and low gravity, blocks of material may be thrown from an impact in a relatively preserved state. But later studies seem to contradict this scenario," Radebaugh says. "They appear to be more like parts of a remnant chain, maybe dissected fluvially or by wind. That's actually in contrast to one possible model from a mountains paper I wrote early on that said there should be crater-ejected mountain blocks – I don't think the evidence is there."

Far to the south, and to the west of Sinlap in the region called Xanadu, rugged mountains appear eroded by river action. Xanadu is a vast area of

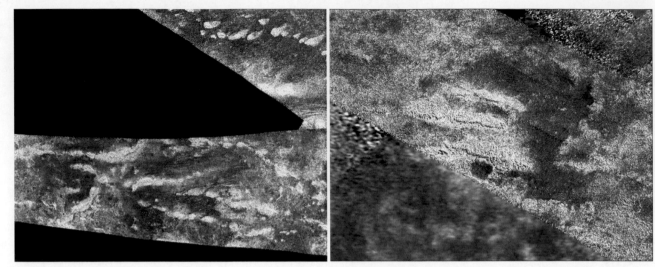

Fig. 7.14 Parallel mountain ranges in the Adiri region (left) and in the northern edge of Xanadu (Images courtesy of NASA/JPL/Caltech/ASI)

mountain peaks crammed together, transected by river valleys that look as if they have carried material fanning out across the plains. Randall Kirk points to "whole canyon systems eaten into flat-topped plateaus, like the desert southwest…very much like Bryce Canyon."

Titan's icy summits are no Alps; they have gentle slopes with low profiles. Elevations range from 120 m to just over 1,000 m. The consistent altitudes are mystifying. Why would all the peaks be so similar in height? It may be that the mountains are all ancient and heavily eroded. Perhaps they overlie a warmer layer of ice that cannot support structures beyond a certain height. Their erosional rates may vary with impurities in the water-ice that makes up Titan's surface. More detailed imaging needs to be done to find out if all the mountains are ancient structures, or if they are in various stages of formation, controlled by an as-yet undetermined cause.

Fig. 7.15 Fault lines cut across wrinkled mountains in two areas on Titan's plains (left). Ejecta thrown from craters like Sinlap was once thought to create mountain ranges (Images courtesy of NASA/JPL/Caltech/ASI)

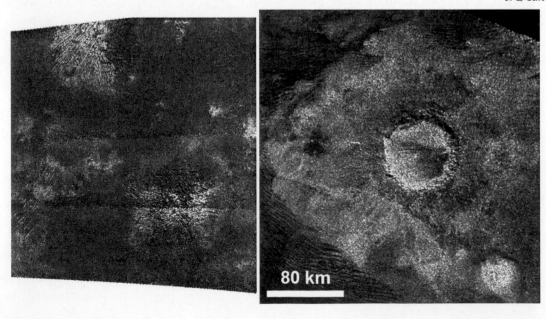

80 km

Titan's highlands differ from the uplifted wrinkle ridges on other planets. They also contrast with the mountain ranges on Venus, which seem to be folds in the surface. No evidence has been found for such folds on Saturn's moon. They do, however, resemble water-eroded mountains on Earth. Radebaugh says this is significant. "On Earth, the dominant land form is the river channel. On other planets, the dominant form is the impact crater. Now we find Titan, where the dominant form is rivers, like Earth." In southern areas, channels cut through terrain with different intensities. Some scour the surface, while others incise canyons with steeper sides. The variation may be due to differences in the density of materials over which they flow.

LAKES AND RIVERS

Titan is the only world besides our own with an active cycle involving rain fed by evaporation from surface lakes and rivers. Its river valleys appear to drain into liquid-filled basins, some as large as the Black Sea. Scientists have postulated liquid methane or ethane on Titan's surface since the late 1960s. After Voyager, the idea of a global ocean was in vogue. But the remote sensing data with Earth-and HST-based observations, combined with the early Cassini data, suggested that Titan was a dry desert world. Standing bodies of liquid remained as nothing but hopeful conjecture until 2006.

In that year, the Image Science Subsystem aboard the Cassini spacecraft observed the first lakes. Images showed a dark expanse near the south pole, now known as Lacus Ontario. The radar-dark lake even looks like the North American Great Lake of the same name. The area was imaged in greater detail by the RADAR instrument. Later, Cassini's radar eyes peered beneath the orange fog to reveal hundreds of small lakes, mostly contained within apparent depressions, in Titan's northern polar regions.

Several lines of evidence supported the interpretation that the areas were lakes rather than simply smooth plains: (1) their morphological similarity with terrestrial lakes and their relationship to river-like features (channels, deltas, etc.); (2) the low radar backscatter, implying surfaces that were very smooth; (3) high methane humidity in the polar regions, which was consistent with computer predictions using atmospheric and climate models; and (4) the radiometric brightness levels of those areas, higher by several degrees than the surrounding terrain, which were consistent with the high emissivity expected for a smooth surface of liquids like methane, ethane, butane and other related substances. Radiometric brightness is different from radar brightness, which shows roughness of the surface. In the case of radiometric brightness, the instrument can tell how much energy is being emitted from the surface. Warm surfaces emit more energy than cool ones, and the lakes appear to emit more energy – they are warmer – than the surrounding land.

Fig. 7.16 *Spanning nearly 500 km across, the magnificent Ligeia Mare is the second largest known body of liquid on Titan. Its radar image is seen here in false color. Inset: The mysterious disappearing "magic island" (Both images courtesy of NASA/JPL-Caltech/ASI/ Cornell)*

Later flybys revealed huge lakes, some similar in scale to terrestrial inland seas. Because of their size, these were soon referred to as maria, Latin for "seas." To date only three have been classified as such. From smallest to largest, they are Punga Mare, Ligeia Mare, and Kraken Mare. Titan's lakes and seas vary greatly in size, from the limits of Cassini's resolution (ranging from 300 to 1,000 m) up to about 400,000 sq. km for Kraken Mare. For comparison, North American's Lake Superior is 82,000 sq. km in extent, and Europe/Asia's Black Sea is 436,400 sq. km.

The large seas found on Titan have rugged coastlines akin to the fjords of Scandinavia. A few of the largest lakes also have some rough features, but the smaller ones are of a very different character. They seem to have mostly circular, oblong or curving shorelines, and their margins are often quite steep. Because of their cliff-like borders, some researchers suggest that the lakes are the result of collapse or melting, much like the rounded lakes caused by melting ice blocks left behind by retreating glaciers on Earth. This type of terrain is called karstic. On Earth, similar regions are very porous and often fractured, with groundwater flowing beneath their surfaces. It may well be that on Titan, these lake regions drain into subsurface methane aquifers that make their way to the lower elevations, eventually feeding the seas. If so, this underground river network may be a major contributor to Titan's atmospheric methane.

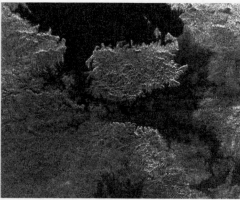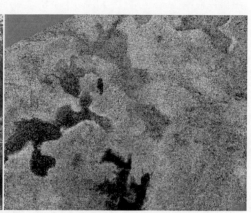

Fig. 7.17 Left: Rugged coastlines reminiscent of Scandinavian fjords line some of Titan's northern lakes. Center: The great island Mayda Insula lies within Titan's largest sea, Kraken Mare, as large as the Black Sea. Right: Lakes of differing radar darkness may be explained by varying depth, or by the presence of methane slush. These rounded lakes may be karstic in nature (Image courtesy of NASA/JPL/Caltech/ASI)

If the lakes have a karstic origin, they provide implications for the chemistry of Titan's lakes and seas. The exact blend of hydrocarbons in the lakes is unknown, but it is likely predominantly a mixture of ethane and methane. Although methane rainfall is thought to be about 100 times greater than that of ethane – which may also be mixed into the precipitation – methane is far more volatile than ethane. Any surface methane would tend to evaporate more quickly. Over time, a standing body of liquid would become enriched with more stable ethane, in which case ethane probably is dominant. But no one knows as of yet just how much liquid is redistributed underground on Titan. If many of the smaller lakes are, in fact, karstic, then it is possible that the entire polar lake district may be underlain by an extensive underground river system, connecting lakes to seas in a web of unseen channels. If the flow is from one general direction, from higher elevation lakes to lower elevation seas, then the seas might have even greater ethane concentrations, as most of the liquid ethane would accumulate in the seas. The smaller lakes may have higher amounts of liquid methane.

It is difficult to tell how full some of Titan's lakes are. In fact, early in the mission, some Titan researchers doubted that Cassini was looking at liquid at all. The problem is that the lakes and seas of Titan are somewhat transparent to the RADAR instrument. The surface of the liquid is smooth, so it acts like a mirror. The incoming radar hits it at a slight angle and reflects away, leaving a dark image. But the liquid methane also lets some of that radar pass through into its depths, like light passing into the clear waters of a lagoon. That radar may penetrate some tens or even hundreds of meters before it fades away or is bounced back by the rough floor of the lake bed. This clearness makes it difficult to tell just where the methane/ethane bath ends and the shoreline begins.

Some subsurface channels were observed within the seas of Titan, and investigators were able to measure depth – or at least general slope – along their lengths. Because rivers flow downhill, it was expected that if these were indeed drowned river valleys, then the channels would get progressively darker downstream, since the spacecraft would observe them through progressively deeper liquid. This, in fact, has been confirmed.

In 2009, VIMS observed a glint of sunlight reflected off the largest of the seas, Kraken Mare, resolving any remaining doubt that Titan's dark areas are, in fact, seas. Researchers are now searching for wave action using Cassini's VIMS in areas where the sunlight might glint off the surface. They can also detect waves in other ways, using radar. Waves of certain size would scatter Cassini's Synthetic Aperture Radar waves, causing a brightening in the image. When the radar is pointed down to carry out topographic mapping, radar-darkened areas can show waves larger than 3 mm. Preliminary analysis has shown several possible cases of wave action on the lakes.

Another phenomenon may indicate wave-caused foam or bubbles on the surface. Cassini glimpsed a bright "island" during a flyby in July of 2013. The bright region in Ligeia Mare was completely missing in imagery from three previous encounters. It disappeared as quickly and mysteriously as it had appeared. It was conspicuously missing on the next flyby just sixteen days later.

Aside from frothy waves, the brightening may be caused by gas bubbling up from the sea floor or icy slush floating on the surface. Methane ice is denser than methane in liquid form, so the slush would need to be a lighter related material, perhaps chains of polyacetylene. Titan experts continue to search images for further sightings.[2]

Adding to the portrait of Titan as a dynamic world, the lakes seem to be changing shape, too. The coastline of Titan's largest southern lake, Ontario Lacus, receded by at least 10 km over a period of 4 years. Several transient lakes nearby disappeared completely in about the same period. The lakes in the north are a different story. Their coasts have been unchanged over a decade of observation. Clearly, they are more stable than those in the southern hemisphere. The northern lakes are more numerous, and their steeper-sided shores may mean that a comparable drop in surface levels may cause undetectable changes in shoreline retreat. It is also possible that the climate in the north is sufficiently different that transport processes are less at this time in Titan's year. The northern ponds may simply be deeper and less prone to evaporation, but most researchers agree that the lakes in the south are more transient and weather-related than those in the north.

Many of the alien lakes and seas on Titan are tied to networks of meandering valleys or outflow regions. Although it is difficult to determine their depths and volume, the vast amounts of methane and ethane in Titan's lakes, seas and rivers are important as both a sink and a buffer within Titan's hydrocarbon cycle, which plays a similar role to the terrestrial hydrological cycle. If, as may be true for Mars, there are extensive reservoirs of liquids under the surface, then the amount of liquid methane on Titan may be much greater – and the transport processes more complex – beneath the surface than it is above.

We would love to know more about the long-term changes as the seasons come and go on this strange world, but the cogs turn slowly in the outer Solar System. By the time Cassini is commanded to enter Saturn's atmosphere at the end of its mission in 2017,[3] it will have observed fewer than two full seasons of weather on the mysterious, fog-enshrouded moon.

2. See "Transient features in a Titan sea" by J. D. Hofgartner, et al., *Nature Geoscience*, published online June 22, 2014.

3. Although Cassini still has maneuvering fuel, flight engineers will eventually command the spacecraft to self-destruct in the atmosphere of Saturn so that it will not crash into either Titan or Enceladus, where there is the possibility of extant life.

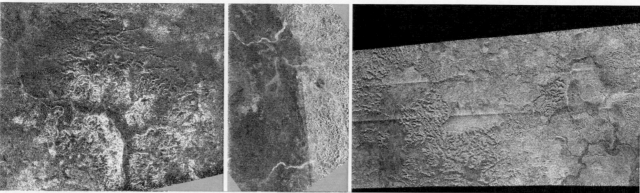

Fig. 7.18 Left: Methane flows have apparently carved elaborate canyons and valleys into this mountainous region. Right: Some rivers at the edge of the rugged Xanadu region may be active today; sinuous valleys and steep canyons point to heavy methane erosion, probably from a combination of rain and sapping, subsurface collapse from underground liquid (Images courtesy of NASA/JPL/Caltech/ASI)

In equatorial regions, Titan is a sort of Arrakis, a dune world washed by rare methane cloudbursts. But in the polar regions, we have seen a different story. There, it may rain seasonally, and methane is a prime ingredient for the climate, both in liquid and gaseous form. Titan's is a different kind of climate from that of Earth, and very different from Mars or Venus. In a sense, Titan is a window into the future, giving us a glimpse of what Earth might be like when the aging Sun becomes brighter and drives the oceans off. Earth will have vast equatorial dune seas, and whatever water is left will migrate to the poles.

Titan has turned out to be a far more complex and compelling world than anyone had predicted. The planet-sized moon baffled, flummoxed, and challenged the greatest minds from the beginning of the first spacecraft encounters. Ralph Lorenz cites its dunes as merely one example:

It was an evolving story. Maybe if we'd been smarter, maybe if we'd had some sand dune experts looking at those first images, they might have twigged sooner, but that's the nature of exploration. You don't know up front what you're going to see. In fact, I had preconditioned myself into not expecting to find sand dunes on Titan at all, because we knew that Titan has methane in its atmosphere, and we knew that the methane will fall down as rain, along with another photochemical product, ethane, which is a liquid at Titan's surface temperatures, so we expected Titan to be damp. When it's damp, you can't move the sand around. We also expected to find seas, and seas trap sand. So the big-picture expectation beforehand was that we wouldn't see sand dunes. We weren't smart enough. Before Cassini arrived, we were thinking of Titan in zero-dimensional terms. 'What is the surface of Titan like?' 'It's like X.' Whereas the reality is, of course, that Titan is a wide world that is diverse. In particular, because it has a long year and a hydrological cycle, there are very strong seasonal and latitudinal effects. Those, as it turns out, conspire to make the atmospheric circulation such as to dry out the low latitudes, and much of the moisture is concentrated at the high latitudes so you find seas around the north pole and sand dunes at the equator. That dichotomy, that profound difference in the landscape as a function of latitude is not something we were smart enough to anticipate.

Perhaps Lorenz's is the most profound lesson of all. Titan is not a uniform ball of ice, nor a chemical laboratory frozen in time, nor even a remote moon to watch and wonder at. It is a varied and dynamic world full of promise and inspiration.

THE CASE FOR VISITING TITAN – A DIGRESSION

Ralph Lorenz, now at APL, trained as an aerospace engineer. He began his career as an intern at ESA, where design decisions were still under consideration for the Huygens Titan probe. Lorenz ended up actually building one of the historic lander's instruments. He has been fascinated by Titan since his teen years, when he read about the strange moon in Carl Sagan's book *Cosmos*. Here is what he said about the possibility of visiting Titan:

Titan has an interesting feature. If you define "Earthlike" as an environment in which an unprotected human would survive longest, Titan is second only to Earth. Put an astronaut on the surface of Venus, and they are crushed with the pressure of 90 times Earth sea-level pressure with an atmosphere at 700 Kelvin. It literally would cook him or her in an instant. On Mars, the low pressure is such that it would suck all the air out of your lungs so that you would pass out within a few seconds. On Titan, you could hold your breath. The atmospheric pressure is just a little bit higher than Earth, so you could probably walk around for a minute until you passed out from lack of oxygen. If you had an oxygen mask – just an oxygen mask without a pressure suit – you'd be fine for rather longer. It's a very cold environment, so you would chill down after some minutes or tens of minutes, but maybe a thick parka or an insulated suit would keep you going for much longer. So you could imagine landing a spacecraft anywhere on Titan and coming out in an environment suit, but the suit wouldn't need to be as elaborate as anywhere else.

It might be interesting to take a sniff of Titan air. It has traces of lots of organic chemicals in it, so it might smell a bit like an oil refinery. In the long term, it would probably be somewhat carcinogenic. We know there are some materials like benzene that carry health warnings on Earth. They might kill you after ten years perhaps, but these are much less injurious in the short term to human health than the extreme toxicity we expect in the Moon dust and the highly corrosive Martian dust, for example. Lunar dust is very angular and abrasive, and the inhalation hazards associated with that kind of stuff is much more immediate and concerning than anything Titan's got.

I have a particular affection for the seas of Titan. In particular, we know that the largest sea, Kraken Mare, has two large basins that are separated by a fairly narrow strait that we've nicknamed the "Throat of Kraken." It's about the same size as the Strait of Gibraltar, about 17 km across. There is a tide in Titan's seas forced by Titan's eccentric orbit around Saturn, so there should be a tidal current that goes one way through these narrow straits at one part of Titan's orbit and then eight days later it goes the other way, a 'tidal race.' It could be that even when there's no wind, those tidal currents may roughen the sea much as they do in tidal races on Earth.

For example, there is Skookumchuck Narrows (Canada), Naruto Strait in Japan, and there's one where I'm from, Corryvreckan in Scotland. You can get a speedboat tour of these tidal straits, and there are whirlpools and very rough sea surfaces, but it's rough in place because it's this tidal current swirling past these bumps and pinnacles on the seabed, spewing off vortices and waves in their wake. You can actually hear this stuff roar.

It would be really cool to either stand on the beach or the cliffs at the edge of this tidal race hearing this – you could hear it on Titan because it has an atmosphere

Fig. 7.19 The tidal race at Corryvreckan in Scotland. Tidal currents interact with the rugged channel floor to set up standing waves and whirlpools. Similar phenomena may occur in Titan's "Throat of Kraken" (Photo © Ralph Lorenz. Used with permission)

that transmits sound very well – hearing the roar and maybe seeing whitecaps on the tidal currents. With a little bit of imagination, and maybe with some polarizing sunglasses to help cut the effect of the haze, at that point on Titan you would be able to see that Saturn was in the sky. Of course, Saturn as seen from Titan will show phases like Earth's Moon.

Flying one of those motorized autogyros over the sand seas would be spectacular, too, seeing these giant dunes stretching from horizon to horizon. Imagining what you would see there is really kind of fun.

Fig. 8.1 Ariel's steep-walled cliffs at the junction of Korrigan and Pixie Chasmas provide a dramatic frame for Uranus. Textures on Ariel's canyon floors hint at previous epochs of cryovolcanic activity (Painting © Michael Carroll)

Chapter 8
Ariel, Miranda and Triton – Moons of Uranus and Neptune

Sir William Herschel discovered the ice giant planet Uranus in the spring of 1781. Within 6 years, he had discovered two of its moons, Titania and Oberon. In 1851, Johann Galle spotted Neptune,[1] and just 17 days later William Lassel found its largest attendant, Triton. It seemed that the outer Solar System was teeming with moons. In fact, it is. The moons of the outer Solar System will provide many targets for human adventurers to study, explore and play upon.

Uranus weighs in with 27 known moons, and Neptune's known total is currently 14. Coupled with the moons of Jupiter (67) and Saturn (62), the total number of known moons in the outer Solar System adds up to 170, and there must be many more lurking out there, especially around Uranus and Neptune.[2] Saturn has shown us the importance of moons as they relate to rings, and the ice giants have rings and ring arcs that are undoubtedly accompanied by unseen shepherd moons. The moons and rings of the ice giants tend to be a low albedo, as if they have been showered in coal dust, so their detection is difficult, to say the least.

In the late summer of 1986, *Voyager 2* arrived at the Uranian system. The craft was old and tired, having been battered by the radiation-filled vacuum of space since its launch nearly a decade earlier. As the craft approached, the humble moons of Uranus began to come into focus. Uranus offered no giant moon the likes of the Galileans or Titan. But researchers had learned their lesson at Jupiter and Saturn. Even the mid-sized moons of Saturn were chock full of surprises for the geologists, and the moons of Uranus promised even more to come.

"The mid-sized satellites aren't terribly well understood," says Elizabeth Turtle, planetary scientist at Johns Hopkins Applied Physics Laboratory, "and all the Uranian satellites fall into that class. We get to Saturn with all its mid-sized satellites, and Enceladus, which is way too small to be active, and has no right to be active, has cryovolcanoes spewing out primarily water. Clearly, we don't understand this class of satellites well. You can explain Enceladus' activity and Mimas' complete lack of activity, but it's not necessarily what one would have predicted."

Armed with the bewildering patterns seen among the Saturnian moons, researchers were reluctant to predict what they might see at Uranus. The surprises continued, and more of the accepted "rules" had to be abandoned, Turtle says. "Then you get to the Uranian system and the smallest [mid-sized] satellite, Miranda, has a completely tortured surface." Ariel, too, has clear evidence of geologically recent activity, while the larger moons appear to be more quiescent and ancient. "It's just not what we would have predicted at all."

As is the case with the gas giants, the satellite system of Uranus may have formed within the ring system or accretion disk in a sort of conveyor belt, migrating toward the planet in a long chain. However, it is estimated that the planet was tipped over early in its formative years, and as we saw in Chap. 4, the tilt of Uranus is a wild 98°.

1. Galle based his discovery on a mathematical prediction by the famous French mathematician Urbain le Verrier.

2. As of this writing, the Cassini spacecraft continues to discover moons at Saturn on a fairly regular basis, especially ones embedded within the ring system.

M. Carroll, *Living Among Giants: Exploring and Settling the Outer Solar System*,
DOI 10.1007/978-3-319-10674-8_8, © Springer International Publishing Switzerland 2015

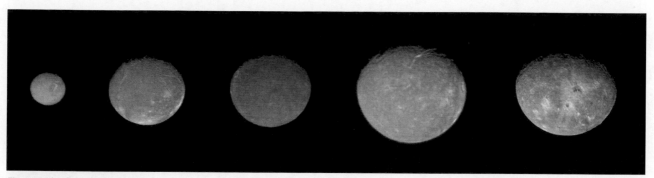

Fig. 8.2 The five major moons of Uranus, seen here to scale, display a remarkable diversity. They are (l to r) Miranda, Ariel, Umbriel, Titania and Oberon (Image courtesy of NASA/JPL)

Computer simulations at the Observatoire de la Cote d'Azur in Nice, France, suggest that several major impacts pushed Uranus into its current tilt, the most severe of any planet. More than one impact was necessary to result in the correct spin of the cloud of debris that led to the moons we see today. As the planet tipped over, the disk of material would have continued to spin around it, trending toward a disk shape in the equatorial plane. This disk then gave rise to a succession of moons that eventually left the ones we see today.

Those five mid-sized moons range in diameter from Miranda's 300 to 1,500 km Titania. The moons all consist of rock and ices of methane, ammonia and water. In an opposite arrangement to the Galilean satellites, the moons tend to increase in density with distance from the planet. They are denser than the mid-sized icy satellites of Saturn, implying that they have more rock compared to their ice component. Miranda is the exception, with a very low density.

Although water-ice covers all five major Uranian moons, it is mixed with a dark material of some kind. Theories range from carbon monoxide and nitrogen flung out of Uranus from its axis-tipping impacts to surface methane darkened by solar radiation. In any case, all the moons of Uranus, from large to tiny, are nearly as dark as coal, as is the ring system.

A BRIEF LOOK AT THE MAJOR MOONS

3. Titania was the Queen of the Fairies in Shakespeare's *A Midsummer Night's Dream.* The moons of Uranus are all named for characters in Shakespeare's plays.

Distant Oberon and Titania[3] are close in size, roughly half the size of Earth's Moon. Oberon, most poorly imaged of all the moons, appears to have a more ancient and battered surface than that of Titania. In the best

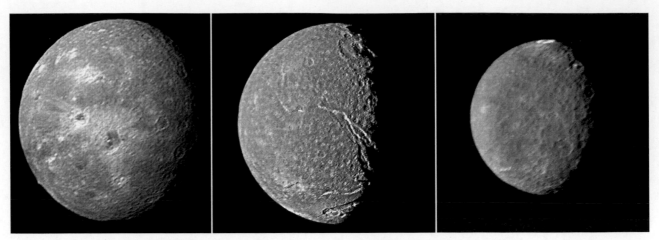

Fig. 8.3 *Voyager's best views of Oberon (left), Titania (center) and Umbriel (not to scale). Note Oberon's 11-km-high mountain on the lower left limb of the moon, possibly the central peak of a crater (Image courtesy of NASA/JPL)*

Voyager images, bright rays of ejecta radiate from several craters, while others have darkened regions on their floors. Titania's landscape is scored by a system of dramatic fault-lined canyons. The valleys appear to be extensional, meaning that the moon either expanded as it froze, or that an outer ice crust contracted and fractured over a more solid core. Both have a larger component of rock within, which may mean that radiogenic heating from core material kept them warm longer than the icy satellites of Saturn, which have more ice and less rock.

Umbriel and Ariel form another pair close in size, each roughly 1,150 km. Although their sizes may be similar, their natures are very different from each other. Umbriel's dark surface appears to be uniformly cratered, with a very primitive face that has changed little from the early days of the moon's formation. This is puzzling, since Saturn's Dione is a near twin in size but has been far more active geologically.

Just next door to dark Umbriel is Ariel. Its dramatic canyons tell a tale of violent upheaval in its past. Finally, tiny Miranda has been the most active of all, with bizarre uplifted chevrons and concentric oval canyons. Of the five major Uranian moons, the two most likely to be visited by humans are the most geologically interesting, and the closest to Uranus itself: Ariel and Miranda.

Fig. 8.4 *Top: Ariel, at left, and Miranda compared to the Mediterranean and Middle East. Bottom: A selection of the exotic mid-sized satellites compared to the smallest of the Galilean satellites, Europa (behind). They are (l to r) Iapetus, Ariel, Enceladus, Miranda and Dione (Montages by the author. Images courtesy of NASA)*

ARIEL

Circling the giant green planet some 175,000 km above its cloudtops, Ariel is the fourth largest of Uranus' five major moons. Despite its small size, it appears to have been nearly the most active, second only to tiny Miranda. Fault-bounded canyons score Ariel's facade into a remarkable parquet surface. Older, cratered terrain tops mesas that rise above bizarre valley floors that have been resurfaced by flows of smooth material. These frozen rivers bow up in the center, and sinuous troughs run through the central sections.

Whatever flooded these valley floors was more dense and viscous than anything seen among Saturn's moons.

Ariel exhibits evidence of resurfacing outside of the canyons, too, perhaps recently in the moon's lifetime. Its ice is the brightest and "freshest" of all the major Uranian satellites. Ariel's face appears to have been resurfaced sometime after the initial bombardment era of our Solar System, whose rain of asteroids and comets tailed out about 3.9 billion years ago. More than twice the diameter of Miranda, Ariel has been peppered by as many objects per square foot in its history, and yet shows a low crater density. The surface lacks many 100-km craters, confirming a geologic age younger than its battered siblings Umbriel and Oberon. Several images show what may be degraded or buried ancient craters, but the well-preserved ones are small and more sparse than cratered terrain of the more geologically quiet moons. Material thrown out from the craters tends to be very bright, and may be fresh ice related to material that has flowed across the surface.

As with satellites of other planets, Uranian moons have impurities that darken in sunlight. If fresh ice erupts and flows across the surface, it will develop a dark crust from the solar radiation, but the light ice underneath will be preserved. The bright haloes around some of Ariel's craters may, in fact, be evidence of ancient, bright flows just under the surface.

APL's Elizabeth Turtle asserts that, "Both Ariel and Titania have regions that look like large rift valleys with flows along their floors that might indicate some kind of cryovolcanism at some point. The pictures are just not good enough to tell. There's definitely been stuff happening there, and more than on any other Saturn satellite except Enceladus."

Past cryovolcanic activity might have been triggered by orbital resonances that no longer affect Ariel today. Multiple study groups[4] are modeling the evolution of the Uranian satellites, and those models indicate that Ariel may have gone through several resonances strong enough to spur volcanism (perhaps with Titania and – in particular – a 3:1 resonance with Umbriel). "People have talked about how there may have been resonances and tidal dissipation [the slowing down of a moon's rotation by gravitational forces of other nearby moons] in the past," says Elizabeth Turtle. "Clearly, something happened at some point. Just by looking at them you know that. Those mesas seem to be older, so the surface broke up and things dropped down and then things flowed along the floor of those rift valleys."

Crater densities on the small moon suggest that Ariel must have been geologically active for an extended period of time, longer than could be explained by radiogenic heating alone. This further bolsters the idea that tidal heating was generated by past resonances from other moons, most likely Umbriel and Titania. Ariel may well have had multiple episodes of activity throughout an early migration of the system, much as the Galileans may have migrated through the Jovian system.

4. See *On the Long-term Dynamical Evolution of the Main Satellites of Uranus* by Karatekin and Noyelles, Royal Observatory of Belgium, published by AAS, 2013.

Fig. 8.5 Left: Flows appear to have obliterated or deformed parts of craters in this high resolution Voyager 2 image of Ariel's Sylph Chasma. Right: A complex network of valleys and troughs surrounds cratered mesas. Note how the valley floors meld into each other, suggesting that they were flooded after tectonic processes formed them. Voyager 2's most detailed images of Ariel have a resolution down to 1.3 km per pixel (Images courtesy of NASA/JPL-Caltech)

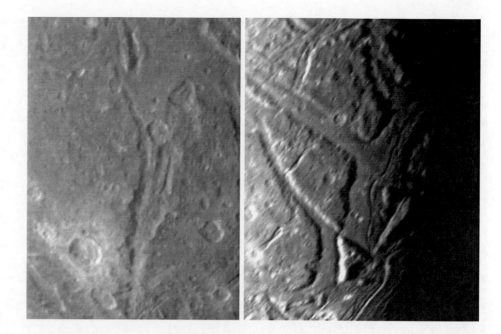

Although researchers have not spotted any specific volcanic sources, they've found plenty of lava-like flows to scrutinize, and the flows contrast with those on the moons of Jupiter and Saturn. There, thin flows tend to form wide plains. On Uranus' Ariel and Miranda, and on Neptune's Triton, flows appear to be hundreds of meters thick. One possible reason may be a difference in composition. Although the ices at Jupiter and Saturn may contain mostly water and ammonia, those at Uranus and Neptune may hold large amounts of comet-like volatiles such as methanol and formaldehyde. The difference in composition would lead to a marked difference in consistency, perhaps resulting in far more viscous cryolavas on the ice giants than on the gas giants.

The problem that continues to plague researchers who search for cryovolcanism among the Uranian satellites is a lack of resolution in the Voyager images. Carolyn Porco, head of Cassini's imaging team, says that Saturn has provided us with a cautionary tale:

> We don't see a shred of evidence for the kind of cryovolcanism that people in the Voyager era thought we had in the Saturnian system. Way back then when we saw the wispy terrain on Dione and Rhea, we thought it was extrusions of ice. So I was fully expecting to see these big hummocky outcroppings of ice on the surface of those moons, like crisscrossing scars. What we found instead was that they're nothing but sub-parallel systems of icy fractures with the freshly cut ice all pointing in the same direction. It's just an optical illusion created by these freshly cracked bands. Even on Enceladus there is no definitive place where ice has come up and extruded out like the magma that's formed the maria on the Moon. We don't see anything like that, and that's been a big result for Cassini. It just does not seem to be present in the Saturn system. It proves how important it is to get up really close.

As luck and celestial mechanics would have it, Voyager did get close to another of Uranus' major moons, Miranda. And that moon turned out to be the strangest one of all.

The Known Satellites of Uranus

	Satellite	Distance from Uranus (x 000 km)	Diameter (km)	Discoverer	Date
1.	Cordelia	50	26	Voyager 2	1986
2.	Ophelia	54	32	Voyager 2	1986
3.	Bianca	59	44	Voyager 2	1986
4.	Cressida	62	66	Voyager 2	1986
5.	Desdemona	63	58	Voyager 2	1986
6.	Juliet	64	84	Voyager 2	1986
7.	Portia	66	110	Voyager 2	1986
8.	Rosalind	70	54	Voyager 2	1986
9.	Cupid (2003U2)	75	12	Showalter	2003
10.	Belinda	75	78	Voyager 2	1986
11.	Perdita	76	80	Voyager 2	1986
12.	Puck	86	154	Voyager 2	1985
13.	Mab (2003U1)	98	16	Showalter	2003
14.	Miranda	130	472	Kuiper	1948
15.	Ariel	191	1158	Lassell	1851
16.	Umbriel	266	1170	Lassell	1851
17.	Titania	436	1578	Herschel	1787
18.	Oberon	583	1522	Herschel	1787
19.	Francisco	1481	12	Holman	2003
20.	Caliban	7169	80	Gladman	1997
21.	Stephano	7948	30	Gladman	1999
22.	Trinculo	8578	10	Holman	2001
23.	Sycorax	12213	160	Nicholson	1997
24.	Margaret	14689	12	Sheppard	2003
25.	Prospero	16568	40	Holman	1999
26.	Setebos	17681	40	Kavelaars	1999
27.	Ferdinand	21000	12	Sheppard	2003

Fig. 8.6

MIRANDA

Miranda is named after the only female character in William Shakespeare's play *The Tempest*. Her name, which means "worthy of admiration," is apt for the most bewildering of Uranus' major satellites. The celestial Miranda displays a host of mysterious geology. Pure serendipity dictated that *Voyager 2's* barnstorming 1989 encounter took it nearest to Miranda of all the Uranian satellites. The craft needed to fly near enough Uranus to be tossed by the planet's gravity outward to its next stop, Neptune. Miranda was in the right place at the right time.

Uranus' diminutive moon seems to have been through a lot. At a diameter of only 470 km – one-seventh that of Earth's own moon – Miranda was remarkably active in its past. The tiny world has baffled the planetary science community ever since *Voyager 2* imaged its southern face. Many mysteries remain, as less than half the satellite has been seen in detail. That hemisphere is wounded in three areas by wrinkled, oblong zones called coronae. These features are unlike anything seen elsewhere in our Solar System. Parallel ridges scar portions of them in great chevrons, while in other regions ice has flowed across the surface, obliterating ancient craters. Towering faults drop precipitously some tens of miles onto rolling terrain.

Why all this activity on such a small moon? Voyager images generated a host of theories involving a nearly-annihilating impact of the satellite as it was still in the process of differentiating. Scientists looked for evidence of such an event in the wild landscapes across the face of this battered little world.

The vast chaotic regions known as coronae mark some of the most remarkable features on Miranda. Spreading up to 300 km across, they consist of concentric ridges and troughs. The impact-scrambling theories suggested that the coronae were dense, carbon-rich portions of a primitive core that was forced to the surface from an immense impact, freezing in place before the material could settle back to the center of the moon.

However, careful study indicates that these regions were more likely formed by uplift and faulting from extensional forces within, perhaps as the moon's ices cooled and expanded. Although some ridges appear to be faults, others resemble volcanic flows. These flows may have been a mix of water and some other material that would act as an antifreeze, perhaps liquid methane or ammonia. The coronae may have formed over plumes of rising material heated by Miranda's core. In this scenario, as material spread out and impinged the surface, it would have triggered cryovolcanism, or at least cryoflooding. Features indicative of flood deposits ooze from fractures at several sites. The parallel ridges may in fact be thick cryolavas that have erupted along fissures. Some sites seem to indicate that the concentric ridges have migrated from the center, covering cratered terrain at their margins. Craters on the corona regions appear fresher than those in the darker, more ancient terrain, and there are fewer of them, indicating a younger age.

Within many of the coronae, ridges intersect in chevron arrangements. The most dramatic of these lies within the center of Inverness Corona. Here, a bright checkmark of ridges spanning 100 km engraves the smoother ridges surrounding it. The more common faults cut across it. One of them, Verona Rupes, wanders far into the cratered

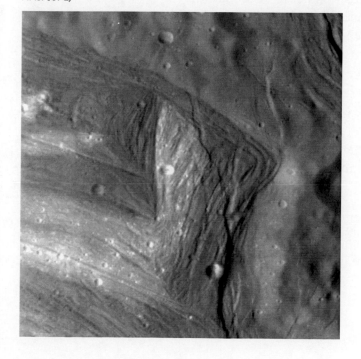

Fig. 8.7 A bright chevron slices across the Inverness Corona. The edge of Elsinore Corona can be seen at upper right (Image courtesy of NASA/JPL)

terrain, rearing up some 7 km, with near-vertical slopes as long as 14 km, creating one of the most spectacular cliffs in the Solar System.

As the great scarp disappears into the darkness of Miranda's terminator, parts of the terrain behind it seem to be collapsing parallel to it. Large sections of the ridge top are precariously scalloped, as if ready to tumble into the abyss at a moment's notice. Perhaps huge avalanche slopes await future explorers, just beyond the sunlit places where Voyager could not peer.

One thing is clear: there's a whole lot of movin' 'n shakin' going on in this little moon, and something must be behind it all. Tidal heating – the same force responsible for the geysers of Enceladus and the volcanoes of Io – may be responsible for internal heat that led to Miranda's jumbled surface. In the past, Miranda interacted gravitationally with Uranus and its nearby sibling moon Umbriel. Umbriel and the small satellite were in a resonance of 3:1 (meaning that Miranda circled Uranus three times for every revolution made by Umbriel).

Fig. 8.8 *Verona Rupes (ending in a cliff at top center), one of the greatest natural wonders of the Solar System, towers nearly 10 km above the surrounding cratered plains (Image courtesy of NASA/JPL)*

Today, Miranda is not in resonance with other moons, and lives a quiet life, silently circling the green ice giant. But it has left us a clue to its peculiar orbital past in the fact that its modern orbit is tilted, or inclined, to the plane of the others, which all orbit Uranus parallel to its equator. Miranda dips up and down at an angle of 4°. Perhaps its internal heat is long gone, leaving only shadows of an earlier epoch in which Miranda migrated through the Uranian system, subjected to gravitational battles that tortured its surface and core.

Like its sibling moons, Miranda has craters down to the limit of Voyager's resolution, but in many places these are wiped out by tectonic or cryovolcanic forces. At one site, a cryovolcanic source may be visible within Elsinore Corona. At the edge of Miranda's terminator, just visible looming from the shadows, is a raised rim broken by what appears to be a flow quite similar to some types of terrestrial lava flows.

Other landforms show us that Miranda's upheavals took place over an extended period, rather than in a single event. Fractures slice across both the coronae and the cratered terrain, ignoring changes in the landscape beneath them. Many are in parallel sets, implying expansion of the ices below.

Fig. 8.9 Possible cryolava flow sites in Elsinore Corona. Insets show details, along with parallel line drawings approximating the raised surface features and, in the case of the feature to the right, possible source and flow trending toward lower right (Image courtesy of NASA/JPL, insets by the author)

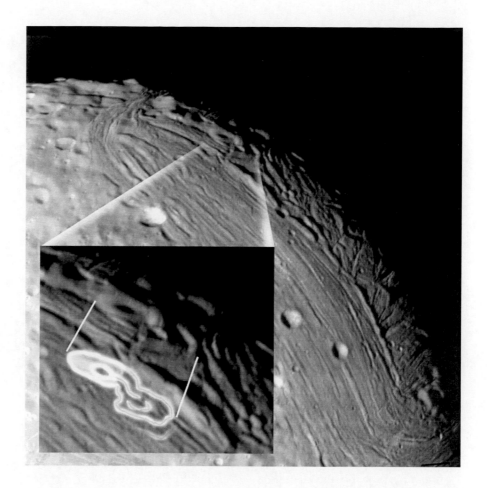

Miranda presents interpreters with the problem of consistency. Of the three major coronae regions, both Elsinore and Arden Corona can be explained well by the impact scramble theory. But the third, Inverness, is complicated by its bright chevron, and its general form makes more sense as a cryovolcanically induced structure. Faults and fractures wander in consistent ways in some regions, and take surprising turns at others. It is likely that Miranda's origins are as complex as its geology is today. How perfect, then, that this little moon be named after the Shakespearian character who declares, "O brave new world!"[5]

Visiting Ariel and Miranda

Elizabeth Turtle, of Johns Hopkins Applied Physics Laboratory, had these thoughts about what it might be like to visit these moons in person:

> I started studying astrophysics in college, and I found that I liked the planets because they are so much more immediate; you can actually send things there, ideally pick up samples, bang on the rocks and see what they're made of. So I moved into the planets.
>
> The Uranian system would be pretty spectacular to be able to observe, from any of the satellites. Uranus isn't as big as Jupiter, but it would still subtend an impressive part of the sky. If you go there near the equinox there would be a lot of atmospheric activity.

5. *The Tempest*, end of Act 1, Scene 5.

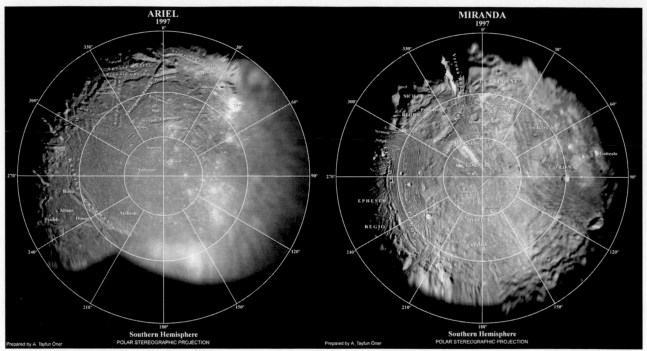

Fig. 8.10 *Polar maps of the Uranian satellites Ariel (l) and Miranda. At the time of the Voyager encounter – the only high resolution data we have – only the southern hemisphere of each moon was illuminated because of Uranus' bizarre tilt. As with all satellite maps, the 0° longitude line points directly toward the planet, while 180° points away (Image courtesy of A. Tayfun Oner. Used with permission)*

I think you'd end up with a lot of extreme sports on Miranda, because the topography is just so outrageous and the gravity is next to nothing. I think you'd have an awful lot of fun with that combination. From just pure, awesome vistas, the topography on Miranda is probably going to win. For sheer, spectacular topography, Miranda is probably the place. But personally, I think Ariel would be a place that would be really fun to wander around on. At Uranus you don't have the whole radiation issue, which makes all the moons easier targets than the Galileans. You could set something up on Oberon or Titania, but Ariel would be the place to go, speaking as a geologist. If you got one of those canyons lined up so you could see through it to the limb, you could see Uranus in the sky. I don't think you could go wrong.

Southwest Research Institute's John Spencer adds:

You'd get the best view of Uranus from Miranda. We don't know the other moons nearly as well, but they are strange and exotic and totally unexpected. On Miranda, you have these huge ice cliffs and there is a lot of bright and dark material and we have no idea what any of it is, so you would have a lot of contrast on the surface. There is mantling that looks a lot like the mantling that occurs on Enceladus from plume fallout, so it makes you wonder if sometime in the past there was some kind of plume activity. There is no sign of it now, and we don't expect it right now because right now there is no tidal heat being dumped into Miranda. Something has certainly heated it in the past, and not that long ago, as some of the areas really don't have that many craters.

TRITON

At the outer edge of our main planetary system, sailing through the melancholy hinterlands of Neptune, circles Triton, the most distant and the coldest moon so far visited by a spacecraft.

The Known Satellites of Neptune

	Satellite	Diameter (km)	Distance from Neptune (x1000km)	Discoverer	Date
1.	Naiad	96x 60x 52	48.2	Voyager 2	1989
2.	Thalassa	108x 100x 52	50	Voyager 2	1989
3.	Despina	180x 148x 128	52.5	Voyager 2	1989
4.	Galatea	204x 184x 144	61.9	Voyager 2	1989
5.	Larissa	216x 204x 168	73.5	Reitsema, Hubbard, Lebofsky and Tholen	1981
6.	S/2004 N1	~18	105.3	Showalter	2013
7.	Proteus	436x 416x 402	117.6	Voyager 2	1989
8.	Triton	2705	354.7	Lassell	1846
9.	Nereid	340 ± 50	5514	Kuiper	1949
10.	Halimede	~62	16611	Holman, Kavelaars, Grav, Fraser, Milisavljevic	2002
11.	Sao	~44	22228	Holman, Kavelaars, Grav, Fraser, Milisavljevic	2002
12.	Laomedeia	~42	23567	Holman, Kavelaars, Grav, Fraser, Milisavljevic	2002
13.	Psamathe	~40	48096	Sheppard, Jewitt, Kleyna	2003
14.	Neso	~60	49285	Holman, Kavelaars, Grav, Fraser, Milisavljevic	2002

Fig. 8.11

As moons go, Triton is large, and orbits in a retrograde direction, opposite the direction of Neptune's spin. If Triton had formed as part of the Neptunian system, physics dictates that its orbit would take it in the opposite direction, revolving *with* the planet and any of its native moons. Another odd characteristic of Triton's orbit is that it is inclined to the equator. Additionally, in comparison to Jupiter, Saturn and Uranus, it orbits in the same relative region – compared to its parent planet – that should be inhabited by a family of major satellites. But there is only one, Proteus. What happened to all the others?

Clues to the answer may lie not only with Triton but with the company it keeps. Voyager's race through the Neptune system revealed six previously undiscovered moons. Proteus is the largest, some 420 km across. Its respectable size gives it membership in the same club as Saturn's Mimas and Enceladus, along with Uranus' Miranda. Like Mimas (but unlike Enceladus and Miranda), Proteus appears to be a heavily cratered, rugged body little changed from the days of its formation. And instead of Proteus being the sibling of a family of major moons, it represents a sort of bookend of a region devoid of large moons.

The other bookend of this region, with a respectable diameter of roughly 350 km, is the unusual Nereid, Neptune's outermost known moon. Nereid has the most eccentric orbit of any planetary satellite yet found. The jumbled orbits of Neptune's system of moons may be the aftermath of a colossal early encounter in which Triton passed near enough to Neptune to be captured. This violent interaction would have left Triton orbiting in its retrograde fashion and would have destroyed or ejected many of the major moons from Neptune's system, leaving the remaining satellites in odd orbits.

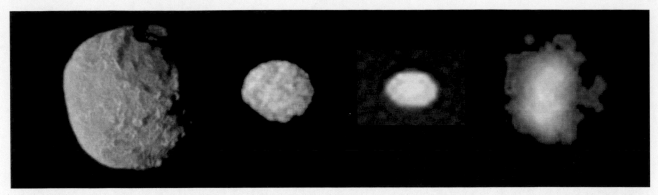

Fig. 8.12 Best Voyager views of (l to r) Proteus, Larissa, Despina and Nereid, roughly to scale. Note that Nereid was imaged from furthest away at lowest resolution (Images courtesy of NASA/JPL)

Triton's orbit eventually circularized, settling into the path it takes today and leaving behind a dearth of other major moons. The small survivors all span less than 220 km in diameter.

Neptune's retinue of moons have been imaged at low resolution, and though their small size implies simplicity and primitive natures, the outer Solar System has taught us that they may well hold a few surprises. But Triton fires the imagination of many a planetary scientist, including William McKinnon. "Triton is a wild and exotic world. At Neptune, the solar winds are weak, so it's a benign environment. You've got a sublimating atmosphere and these miles-high geysers. I'd love to have boots on that one." Triton's exotic nature may stem, in part, from a global resurfacing after its wayward beginnings. When the interloper entered the Neptunian system, it may have been a heavily cratered ice-rock ball. Gravitational forces during its capture would have heated its surface to the melting point. Any rock within the satellite would have had ample chance to settle to the center in the process of accretion, and *Voyager 2* gravitational studies seem to bear this out. But the ice above this rocky core would have completely melted into an ocean as Neptune's gravity forced Triton's eccentric orbit into a more circular one. Eventually, the water became an ice crust, but processes are still at work today, sculpting the moon in puzzling and outlandish ways.

Frustratingly, only a third of the moon has been imaged in moderate to high resolution (400 m per pixel). Within that tantalizing slice of globe, researchers have been able to identify only 15 impact craters. This lack of craters underlines just how active and geologically young the surface of Triton is even now.

Common to much of Triton's real estate are the peculiar dimpled flatlands called cantaloupe terrain. John Spencer of the Southwest Research Institute has been studying it, along with other phenomena on Triton, in association with dwarf planets in the outer Solar System's Kuiper Belt:

> The cantaloupe terrain is certainly distinctive. It almost looks like stacks of plates arranged on a table with their upturned rims. It's a strange, broken-up landscape. We have this one detailed picture, the last but one high resolution picture that Voyager took, where you have an oblique view across the cantaloupe terrain, and

Fig. 8.13 *The mysterious cantaloupe terrain may be among the most ancient regions on the surface of Triton (Image courtesy of NASA/JPL)*

Fig. 8.14 *Ruach Planitia, with its stair-step cliffs and volcano-like central pit, may be an ancient cryovolcano. At left is the original Voyager image, while at right is a computer-generated oblique view with exaggerated vertical scale for clarity (Image courtesy of NASA/JPL/ Caltech)*

you get some idea of what that landscape would look like if you were on the surface. There are ridges cutting through, but the ridges are all warped, like someone attacked them with a blowtorch. There's a lot of stuff going on there that we have no understanding of. The ridges on Europa look very straight and pristine, as if they are from yesterday. On Triton, everything is slumped down and warped and subdued and buried in weird ways.

Judging by its relationship to some of the other geologic features, cantaloupe terrain may represent the most ancient of the regions on Triton. Its plates, called cavi, are 25–30 km across, and are crossed by interconnected ridged valleys. The processes that led to the cantaloupe terrain may involve some as yet unknown erosional activity, or the territory may be the result of ices evaporating away after they form. In many places, low cliffs edge the plate-like plains.

In other areas, cliffs stair-step down to smooth plains. Some of these plains have been flooded, perhaps by cryovolcanic flows. One in particular, Ruach Planitia, appears to have frozen waves across its floor. At its center lies a collapsed pit that shares characteristics with volcanic calderas.

A vast calligraphy of bright salmon-colored material stretches across the southern hemisphere of Triton in the Voyager encounter images. At the time of encounter, Triton was approaching its summer solstice in the south. The rosy material appears to be nitrogen frozen to the surface as ice. At the ice edge, complex lake-like structures rest within bright haloes of material. These are called guttae, or lacus (Latin for "lakes"). They appear to be left behind by Triton's retreating polar cap, explains SwRI's John Spencer. "The lacus areas remind me of melted wax. It looks like liquids have flowed on the surface, but it doesn't look like flowing water

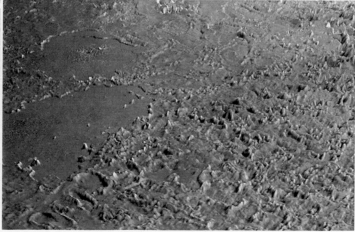

or flowing lava; it looks more like wax. I'm sure it's probably some kind of cryovolcanic ice, but it's all very alien. The lacus areas have a weird, scalloped edge to them. It looks like whatever was liquid in there was melting and collapsing the edges, eating away at them and eroding them back. There's a great deal happening there."

Triton's most remarkable features of all, however, are its active eruptions. *Voyager 2* skimmed Triton's south pole on August 25, 1989, taking stereo images that showed two dark, tall plumes, reaching about 8 km above the surface and leaving trails for about 150 km. Other images of Triton's southern polar region revealed more than 100 dark, streaky deposits, pointing preferentially northeast, away from the southern polar ice cap. The streaks stretch from tens to hundreds of kilometers across the surface, implying that plume activity must be fairly common. Researchers have no idea how widespread this volcanic activity is on Triton, as the northern polar regions were in darkness during the Voyager flyby. It does not seem to have occurred at lower latitudes in the icy plains and cantaloupe terrain, but is restricted to the icy polar cap.

Fig. 8.15 Lake-like features called guttae are actually frozen solid ponds of unknown darkened material. They are unique in the Solar System. The upside-down mushroom-shaped gutta in the foreground is about 100 km across (Image courtesy of NASA/JPL)

Triton is very, very cold, −235 °C at the surface. This is well below the freezing point of nitrogen, the material that makes up the southern polar cap. Ground-based telescopes have identified both nitrogen and methane in Triton's surface spectrum, and carbon monoxide and carbon dioxide have been detected in lesser amounts. Triton's thin atmosphere consists largely of nitrogen. The atmosphere transports nitrogen ice from pole to pole every Triton year, as the polar cap sublimates, turning into a gas and freezing again at the opposite, cooling pole. This migration of nitrogen from one pole to the other keeps the surface temperature nearly the same everywhere.

In this strange, cryogenic world, how does volcanism happen? Cryovolcanism on Triton is very different from cryovolcanism on the satellites of Jupiter and Saturn. It is thought that the plumes are similar to geysers, but caused by solar sublimation – melting directly from ice to vapor – of the nitrogen polar cap. The active plumes were located close to the area where the Sun was most directly overhead.

One model explains the plumes as the result of sunlight passing through nitrogen ice, which is very clear. The Sun's energy becomes heat in a solid-ice version of the greenhouse effect. Heat builds up inside, melting the nitrogen ice into nitrogen gas. The vapor pressure builds up and finally explodes into the near-vacuum of Triton's environment.

Fig. 8.16 The highest resolution mosaic currently in existence of Neptune's exotic moon Triton. The equator runs roughly horizontally across the center of the image set. The pink material is nitrogen ice (Image courtesy of NASA/JPL)

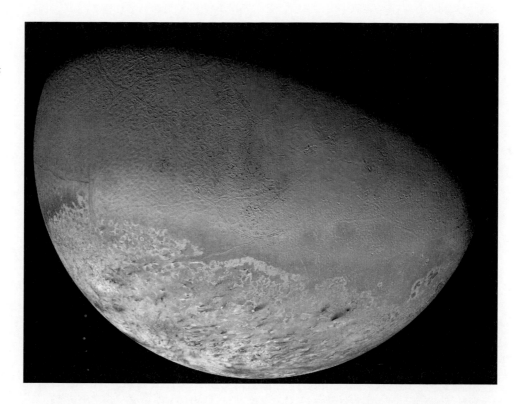

If this model is correct, cryovolcanism on Triton is a side effect of sunlight, rather than an internally driven phenomenon. John Spencer adds, "What's interesting is that this idea was proposed back in 1990 by Bob Brown and Randy Kirk, and since then we've found that this exact mechanism operates on the Martian polar caps. We have no idea if it works on Triton or not, but it really does work on the Martian polar caps, and you get these spider-like arachnoid features that seem to be due to exactly this mechanism. The Sun is penetrating the seasonal carbon dioxide ice and evaporating it, and it's coming up through cracks in the ice and making jets. I always like the idea that every theory that's been proposed for the Solar System is true, but often not for the object it was proposed for."

Outer planet experts got a provocative glance at Neptune's wondrous moon during Voyager's flyby over two decades ago. Since then, Triton's seasons have progressed from mid-summer in the south to late southern autumn. (Uranus and Triton experienced equinox in 2007, where the Sun was directly over the equator; it is now making its way north.)

Twenty-five years after Voyager, observers have learned more about the moon, but not in the same detail as before. Rather, Earth-based telescopes have detected dramatic changes in Triton's color. From a fairly neutral beige, some researchers have watched the satellite shift to hues of a striking reddish-brown. Assessing its color is difficult because of its proximity to Neptune itself, and there are several alternative explanations involving the color change actually being an artifact of the instruments, but the observations have been independently carried out by several teams.

If Triton is changing color globally, the color shift might be due to geyser activity at the pole infusing the thin atmosphere with reddish smog. Alternatively, it might be the result of some kind of super-plume eruption, where large areas are being resurfaced. Spencer prefers the former idea. "I favor the 'smog' hypothesis (maybe due to a surge in geyser activity, which could have a global effect even if the geysers were in the southern hemisphere) because it's the easiest way to change the color of the whole of Triton, as is probably required to produce a large change of the average global color seen from Earth." Researchers do not know where on the surface the seasonal changes are occurring, but the fact that they vary as Triton rotates suggests that they are probably not near the pole but rather equatorial. These changes may be due to gradual evaporation of frost, rather than directly from geyser activity.

Triton is extremely rich chemically. Scientists have obtained very detailed spectra in the infrared, and its light betrays the presence of methane, nitrogen, carbon dioxide, carbon monoxide and water. As the moon rotates, they see that the chemistry changes as different hemispheres come into view. But the frustration is that Voyager had no advanced spectrometer aboard. The variety of texture and color on Triton's surface has no quantifiable link to the spectra that modern instruments are providing today. Because of telescopic resolution, the spectra are averaged together in such a way that it is difficult to tell one part of the moon from another. The spectra must then be correlated with the colors that Voyager saw, which might have changed in the last 25 years. It's a tricky business, but researchers continue to tease data from the difficult observations.

With all of her time at the telescope, outer planets expert Heidi Hammel is used to difficult observations. She finds Triton very compelling. "In its own right, it's an amazing moon. It's dynamic. It's young. Things are happening. Whatever we see there will almost certainly be different than what Voyager saw. Also, it's a virtual twin to Pluto, a fraternal twin. It's a Kuiper Belt object that was captured by Neptune. We know that because it's in a retrograde orbit. You take Pluto and capture it by Neptune and melt it and turn it inside out in the process." She adds: "We're sending a spacecraft past Pluto in a couple years so we'll have a fabulous comparative set."

Still thinking of future missions to the Neptune system, both robotic and human, Hammel points out another important feature of Neptune's largest satellite. "Triton is a big moon, so you can use the gravity of Neptune and Triton to tweak your orbit. The biggest moons of Uranus are not big enough to do that." Triton, then, may provide future explorers with a gravitational lynchpin with which to navigate the wonders of this farthest of the giant worlds. And Triton itself entices as a destination. Its stunning geology, spectacular cryovolcanism, and strategic position within the ice giants may make Triton a key player in future human expeditions.

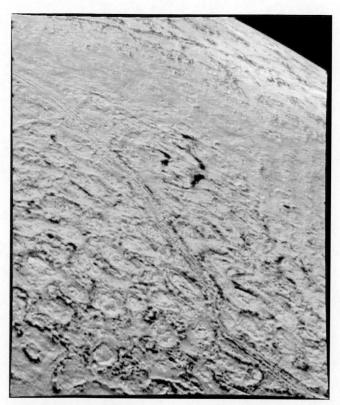

Fig. 8.17 Neptune's largest moon Triton exhibits a variety of landscapes not observed in the rest of the Solar System, including the mystifying "cantaloupe terrain" seen here in the foreground, lower portion of the image (Image courtesy of NASA/JPL)

Visiting Triton

John Spencer, of the Southwest Research Institute, had this to say about us exploring Triton up close:

When we first got the pictures of Triton in 1989, I was there at 3 am watching them come in. I thought, 'This is a world that is as amazing as Mars in terms of the complexity of what's going on there.' It's like you saw Mars for one hour and that's all you'll ever see in your lifetime. There is very little that we see on the surface that we understand. The surface is more chemically complex than any other solid surface we know of in the Solar System. It has an atmosphere, and it has seasons, and these plumes–whatever the heck those are – so it's just a really rich environment.

It would be a fabulous place to explore. I'd land near one of the plumes. From the places where those plumes are, you'd get a great view of Neptune. Unlike the big moons in the Jupiter or Saturn system, where you're sedately going around the equator and you always see the equatorial view, on Triton you've got this real roller-coaster ride. You're going high above the north pole and then swooping down through the equator and going over the south pole and coming back up again, and you do this every six days. The view of Neptune is going to change very dramatically as you go around that cycle. You can see the rings when the geometry is right. They're always changing, too.

We know from comparisons of Voyager images to Hubble images that those rings are evolving quite rapidly. They're clumpy, so you could watch the clumps in the rings going around. It would be quite the view. There may be active cryovolcanoes on the surface. We don't really know what's powering that activity, but it has certainly been occurring recently and may well be happening right now. There are a lot of bizarre landforms like nothing we've seen anywhere else in the Solar System. It's kind of flat. There are cliffs, but they are fairly low, and there are no high mountains, but it would be an amazing place to wander around and see what some of this stuff really is. Of course you'd want to check out the geysers if they were still active. We don't know how long they last. I imagine there are fractures on the surface with these high speed jets coming out of them from stuff that's evaporating from down below, going 5 or 8 miles into the sky, with these long smoke plumes trailing away from them. You won't see that anywhere else.

I think the sky would be black when you look straight up, but we know from Voyager, where we actually have pictures of haze layers near the horizon, that if you look down low you will see – at least some of the time – wispy layers of haze that tell you there is an atmosphere there. If you were under one of those plumes, you'd see that haze streaming overhead, and you'd be seeing a visible atmosphere overhead, because those are dense enough that they would block out the stars, and probably block out your view of Neptune if you were in the right place. That would be quite something.

Part III
A New Frontier

Fig. 9.1 After decades of robotic exploration, humans prepare to follow into the subsurface ocean of Enceladus. A pressurized dome on the floor of a "tiger stripe" valley prevents violent boiling of the seawater into the vacuum of space. The dome and its airlock (right) have been sealed to the surface with a slurry of rock-hard ice
(Painting © Michael Carroll)

Chapter 9

Technology and Living Among the Giants

They came across the emptiness in a flotilla of vessels, with the stars to guide them. Each ship stretched to a length of 19 m and spanned just 5 m across. And yet they came. They braved towering waves, violent winds and blistering heat. Eventually, they were successful, surviving to establish the first civilization within the Hawaiian Island archipelago.

The first explorers to the outer Solar System went at the bidding of the engineers and scientists who built them. They sailed through the inky seas of the void beginning in the early 1970s. In place of bamboo mats, the intrepid explorers cocooned in gold and black Mylar® blanketing. They traded conch-shell horns for great white antennas, wooden oars for hydrazine rockets. Like the early Polynesian mariners, they sent back accounts of the new lands out there, lands of cooled lava, canyons and mountains, and mysterious deep oceans of water.

Who will follow them? Will future human explorers pursue the pattern of their Polynesian forerunners, settling the already surveyed shores of the cosmic sea? Where will we be headed three centuries after the first robotic travelers blazed that path?

In fact, there are strong parallels between the early Polynesian diaspora and any future human migration to space. The Polynesians radiated from the Society, Tahiti and Marquesas islands over 1,000 years ago. Their journeys began primarily as voyages of exploration rather than settlement.

Archaeological evidence from the island of Hawai'i suggests that the first landfalls were made by expeditions from the Marquesas around A.D. 800. Pearl, bone and shell artifacts at early sites are of Marquesan design. The area's wind currents bear this out. If Marquesan canoes sailed northward, using the North Star as reference (the only stationary star in the sky), trade winds would have carried them 15° to 20° of longitude to the west. According to Dr. Richard Grigg at the University of Hawai'i, this would have been "just westerly enough to land somewhere in the high islands of Hawai'i."[1]

Three centuries later, between A.D. 1100 and 1300, the Polynesians set sail with something different in mind. Their double-hulled canoes made landfall on Hawai'i again, bringing with them some 30 varieties of fruit and medicinal plants, pigs, dogs, chickens and rats.[2] This was not an accidental landfall but a carefully planned expedition. Polynesian explorers had come before; the new Hawaiians knew where they were going, and they planned to stay.

"If you want to settle, you want to bring your pigs and your chickens and your women," says science artist and scholar Jon Lomberg. "That's why the old explanation of, 'Oh, some fishermen were out and they just got blown off course to another island,' doesn't work. You don't take chickens and pigs and breadfruit saplings if you're going fishing. These were intentional voyages of colonization. The canoe was the island. The island was the canoe. It wasn't that the people were sailing. It was actually that part of

1. For an excellent overview of the Polynesian diaspora throughout the southern Pacific, see *In the Beginning, Archipelago: The Origin and Discovery of the Hawaiian Islands* by Dr. Richard W. Grigg (© Island Heritage Publishing, 2012).

2. There is debate among anthropologists and archaeologists as to whether the rats were purposely brought along as quickly reproducing sources of protein, or if they were merely doing what comes naturally to rats – stowing away. Other possible interlopers may have included geckos and skinks. To this day, Hawaii has no snakes.

M. Carroll, *Living Among Giants: Exploring and Settling the Outer Solar System*,
DOI 10.1007/978-3-319-10674-8_9, © Springer International Publishing Switzerland 2015

Fig. 9.2 Breadfruit grows independent of seasonal change (© Hans Hillewaert, via Wikipedia Commons. http://upload.wikimedia.org/ wikipedia/commons/5/57/ Artocarpus_altilis_(fruit).jpg

the island, soil from the island, plants from the island, and of course insects whether you wanted them or not. They also knew: 'Yeah, it's possible to come back some-day.' But you were gone for a long time. This was not a weekend trip."

If people wanted to learn long-distance navigation, the Pacific was the place to do it. The islands of the Philippines and Indonesia are spaced close enough to move easily from one to another. Navigation was not a problem; voyagers could see their destination. Polynesian cultures spent generations developing ship-building technologies and seamanship skills. Eventually, they began to take longer trips. "It's kind of like training wheels," says Lomberg.

As you move eastward through the Pacific, the island groups get spaced wider and wider. But by the time your culture gets there you've spent the last 25 generations learning about building and sailing canoes, what you had to bring with you, how to carry water. So you're setting out a voyag-ing canoe to settle another island that might be a couple month's sail away; once you get out to Hawai'i the distances get extremely large. Except for the islands in the chain, they're several thousand miles away from anywhere. So to do long-range voyages, you had to fit out the canoe. It's like the way NASA is at the apex of our culture in engineering and materials and science. These canoes and navigation techniques were the same for them.

As the Polynesians learned, incrementally, how to make longer and longer voyages, so spacefaring humans have built vehicles to withstand suborbital flights, followed by trips to low Earth orbit. Astronauts and cos-monauts ventured higher and farther, and built large orbital structures such as the MIR. These journeys culminated in translunar treks and lunar landings. Aerospace engineers are learning technologies that will enable humans to travel, eventually, into the outer Solar System.

MADAME PELE'S BOUNTY MEETS IO'S WILDERNESS – A DIGRESSION

Taking stock of the ways of ancient Pacific settlers, modern designers are proposing similar menus for long duration space voyages – and even for settlers of other worlds. For example, the Four Frontiers Corporation sug-gests bamboo as a fast-growing, strong material in future off-world green-houses. The fibrous material can be used in construction and as a quickly reproducing, robust oxygen-maker. Breadfruit trees, another Hawaiian standby, produce highly nutritious fruit in large volumes. The trees have the added advantage of bearing fruit year-round, so they are not vulnera-ble to seasonal change.

Fig. 9.3 The Hokule'a, a modern reconstruction of the ancient Polynesian double-hulled canoe. Craft like this populated the entire south Pacific region over the course of a millennium (©Todd Carle, Hawaii Kai Boat Club)

The deep seas surrounding the Hawaiian archipelago are dwarfed by those at Europa, Ganymede, or perhaps even Enceladus, but to travelers in an open craft, the depths commanded respect nevertheless. The island peoples lived with the sea, knew its nuances and moods, and were keenly aware of the toothy beasts, treacherous currents and other dangers lying beneath. Canoes from Tahiti, Fiji, Samoa and the Marquesas traveled across literally thousands of miles.

The great explorer Captain James Cook famously observed, "How could a stone-age people have navigated and explored a third of the Earth's surface without instruments and charts?" The answer, in part, was that they didn't need them. True, ancient navigators sometimes charted the stars using mats of twine and sticks, but they had other techniques at their disposal. They watched ripples and swells of the waves to catch telltale signs of distant shores. They looked for billowing cumulus clouds that tended to stack up above islands, and they kept watch for the green glow of forest and atoll reflected on the base of those clouds.

From generation to generation, they passed down knowledge of the comings and goings of the trade winds, and the times when those winds reversed direction or brought storms. Those who would dare sail into the unknown memorized the relative locations where the stars, Sun and Moon rose and set with the march of the seasons. Even the birds gave them clues; migrating flocks hinted at global directions, and shore birds went to sea in the morning and returned to land at night. Their travels, too, pointed the way to distant coastlines.

"The Hawaiians didn't have a written language because there was no paper in the Pacific to make it out of," says Jon Lomberg, "so there were no written languages in the Pacific." He continues:

> But you can write on your body. That's why they invented tattooing. 'Tattoo' is a Tahitian word. And that's also why they had the petroglyphs. Charts weren't practical, so instead of charts they had chants. The chants were for all the passages of life. There were birth chants and death chants and wedding chants and all these things. But there were also navigational chants that would say: 'If you want to go from Samoa to Guam, here's when you sail and here are the stars you are sailing towards.' It's the rising and setting stars that are the most useful for navigation. Just like for us, it's easier to remember a song or prose. They would learn these chants and would pass them on.

FOLLOWING TRAILS

A lot can happen in three centuries, or even in one. In the span of less than 100 years, the Industrial Revolution changed economies, landscapes and life in Europe completely, from high society to the local mining communities. The mighty Soviet Union, a nation whose military might and communist philosophy influenced half the world, came and went in the course of just over 70 years. The twentieth century opened with the invention of powered flight; scarcely two thirds of the way through, humans were voyaging to the Moon. We've gone from boiled cabbage to microwaved linguini. These advances are not reserved for the royal or privileged, but are for everyone, for the person on the street, in the cubicle and on the assembly line. No matter how advanced, technology eventually spreads to humankind at large. Is it too great a jump to see our inventions carrying us to the moons of Saturn or Neptune or even Uranus within the next century?

Historically, the main influences that have triggered or defined exploration are technology, science and research, and geopolitical considerations. Of these, the latter has been the most prominent. Columbus did not come to the Americas to find out what kind of plants were growing there. He came to find a more efficient trade route, and he came at the sponsorship of a politically and economically motivated government. The Apollo project revolutionized both science and technology, but its motivations were primarily geopolitical, a dash to beat the Soviet Union in a technological race for national prestige.

Who will be the geopolitical players of the outer planets? China has been a massive world influence for millennia, but what it will be like in a century is anyone's guess. The power of Europe and the United States ebbs

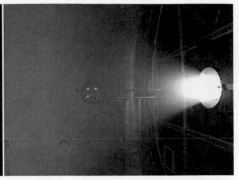

Fig. 9.4 Left: Two kinds of conventional rocket engines are used to propel the space shuttle Atlantis during launch: solid rocket boosters on the left and three space shuttle main engines, powered by more efficient liquid fuel, on the right. (Image courtesy of NASA/JSC.) Center: The blue glow of ions illuminates an engine test for the Deep Space 1 solar-powered ion propulsion mission. (NASA/JPL). Right: Ad Astra Corporation's advanced VASIMR engine combines the best of multiple propulsions (© Ad Astra Rocket Company. All rights reserved. Used with permission)

and flows, while nations such as Brazil are growing in economic and political power. Russia may fade into a backwater has-been state or resurrect itself into a new union of states, perhaps fundamentally different from the space-faring Soviet Union of past days. India today is a world player, with the world's 11th largest gross national product, the second-largest population, and a vibrant indigenous space program. India may well be one of the driving political forces to come, though some analysts say the supercontinent shows signs of a weakening economy and flagging influence.

One thing is certain. When it comes to the future of global geopolitics, nothing is certain. But technology is a different story. History has seen a stair-step progression of technology, and it seems a good bet that our clever engineers and inventors will enable us to venture – in person – into the outer Solar System, perhaps within the next century.

To the early twenty-first century mind, the advances of the next few decades will be akin to science fiction. Laboratories across the world are carrying out research in advanced propulsion ("Ahead warp factor one."), human hibernation ("Get your paws off me, you damned dirty ape!"), food production in microgravity ("You are *not* using those things in my forest."), and human-technological interaction on long-term voyages ("I'm sorry, Dave; I'm afraid I can't do that.").[3] Research continues to advance on many fronts.

PROPULSION: JUST GETTING THERE

The most daunting aspect of human exploration of the giant worlds is, very simply, the distance involved. Our closest target, Jupiter, is 588 million km at its closest pass, but we cannot travel so short a distance directly. Traveling to another world is a game of cat and mouse, with an Earthship coasting along an arc stretching across the Solar System to a point in front of the target planet. In effect, the spacecraft must place itself in a position for the planet to catch up to it.[4]

Further, crossing millions of kilometers of the void involves great spans of time, and for human crews, those spans translate into long-term

3. References are from *Star Trek*, original series ca. 1966–1969; *Planet of the Apes*, 1968; *Silent Running*, 1972; and *2001: A Space Odyssey*, 1968.

4. Assuming a direct flight, but with large craft, planetary encounters are planned using slingshots to gain enough velocity to get to the outer system. The Jupiter-bound Galileo mission made use of multiple flybys for gravitational assists, clocking an incredible 3.86 billion km before reaching its goal.

radiation exposure, subjection to extended microgravity environments, and the consuming of vast quantities of supplies. Says space shuttle veteran Alvin Drew, "The longer you're out there, the bigger your logistics problem is going to be – having to have recyclable water, regenerative food supplies, reconstituting your atmosphere." One solution is to trim down the travel time by developing more powerful or efficient propulsion.

Engineers, scientists, experimenters and science fiction writers have been exploring many types of propulsion. In addition to the familiar chemical propulsion used by most spacecraft today, experiments have been carried out using nuclear propulsion, electric ion drives and solar sailing. To date, only two kinds of propulsion have been used to send spacecraft on planetary missions, conventional chemical rocket engines and solar electric, or ion drives.

Chemical Rockets

Conventional, or chemical rocket engines, the kind that lofted astronauts to the Moon and to orbit aboard Soyuz and shuttles, carry fuel and oxidizer, each stored separately. When the fuel and oxidizer mix, they create a controlled (ideally) explosion that is sent out the back end of the rocket through the engine bell. In effect, a rocket moves by ejecting mass in the opposite direction of its flight. Rocket fuel takes up a lot of physical space, and it is heavy. The higher the temperature of the exhaust, the faster the vehicle goes. The amount of propellant that a rocket engine uses to create a given amount of force is called its specific impulse. Solid rocket boosters – such as the SRBs strapped to the sides of the shuttle's main tank – have a specific impulse, measured in seconds, of about 250. The far more efficient liquid-fueled space shuttle main engines have the highest specific impulse of any conventional engine, at 455 s. With conventional rockets, most of the fuel is spent with a burst of acceleration at the beginning of flight, as storing the fuel is difficult.

But once out of Earth's atmosphere, modern deep space probes are more often using an efficient form of propulsion that can provide steady thrust over the long distances involved in outer planets exploration – ion or electric propulsion. Instead of heating chemicals, ion engines heat gas into a plasma.

Ion Drives

An ion drive can go faster on a lot less fuel. The specific impulse of an ion engine is very high, on the order of 3,000 s, but its acceleration is gradual. Unlike conventional engines, the ion drive can run over long periods of time. Geostationary satellites regularly use ion thrust to remain in place. The asteroid-comet flyby mission Deep Space 1 used an ion drive for long durations over its 2-year mission, and the Dawn mission used a xenon-gas ion drive to carry it on a looping trajectory to orbit the asteroid Vesta.

It then departed for another orbital study of the largest asteroid, Ceres. At one point, Dawn fired its ion thrusters continuously for 270 days. Dawn's ion drive is capable of accelerating from standstill to 96 kmph in 4 days. This compares to a conventional rocket, which leaves the ground and reaches an orbital speed of 27,360 kmph in less than 10 min.

Ion drives use electricity to accelerate propellant out of the back of the engine. Solar electric propulsion takes advantage of magnetism and electricity to push a ship through space. Electricity, generated by the ship's solar panels, delivers a positive electrical charge to atoms inside the chamber. Magnetic fields pull the atoms toward the back of the ship, where they are pushed by an opposite magnetic field out of the ship. This steady flow of ions streaming out of the spacecraft provides thrust. Electric thrusters typically use much less propellant than chemical rockets because they have a higher exhaust speed. The thrust is much weaker than chemical propellant, but the engine can be run for months rather than minutes.

Thermal Nuclear Rockets

A thermal nuclear rocket uses the heat from a nuclear fission (or eventually fusion) reactor to accelerate its fuel and create thrust. Unlike solar-powered ion or plasma engines (such as those used by the Dawn spacecraft), thermal nuclear engines perform largely the same job as chemical rockets, but they do so using only half the fuel.

Early U. S. experiments with nuclear propulsion culminated in the NERVA (Nuclear Engine for Rocket Vehicle Application), a nuclear thermal rocket engine program that progressed steadily for nearly two decades. The goal was to create a reliable engine for space exploration, and at the end of 1968, engineers confirmed that the latest version of the NERVA engine, the NRX/XE, met all the requirements for a manned Mars mission. The NERVA engines were built and tested with flight-certified components.

Just as the engines were declared ready for integration into an actual spacecraft, however, the Nixon administration canceled much of the U. S. space program, including all work related to manned Mars missions. NERVA was an incredibly successful program. Most of today's designs that incorporate nuclear thermal rockets are derived from NERVA technology.

A more modern version of nuclear propulsion under development is the Nuclear Cryogenic Propulsion Stage, or NCPS. Since the design of NERVA engines, new, lighter materials have become available that can perform at higher temperatures and pressures. Current plans call for an engine complex with a fueled weight of 40 t. For safety, the craft would be launched before the reactor ever is enabled to become critical. In the event of a launch catastrophe, the only radioactive contaminant would be roughly 100–150 kg of uranium fuel. If used as an upper stage, the NCPS would be brought on line only after successfully reaching space. In the case of deep-space probes, the NCPS would not be initiated until the craft is ready to leave orbit.

Fig. 9.5 The IKAROS solar sail, launched by Japan's JAXA space agency, flew to within 80,000 km of Venus in 2010 (Model courtesy Pavel Hrdlička, Wikipedia. http://en.wikipedia.org/wiki/IKAROS#mediaviewer/File:IKAROS_IAC_2010.jpg)

The next step in propulsion is the dream of clean fusion-driven nuclear drives. NASA has funded several early-stage fusion concepts. Some estimates suggest that a fusion rocket could make it possible to get astronauts to Mars in as little as 30 days. Transit times to Saturn would be cut to several months. The problem, of course, is that operational fusion reactors have not yet been developed here on Earth. Thus far, experimental fusion reactors generate as much energy as they take in, despite decades of research and development. But many propulsion experts declare that fusion's day will come, and it may come in the very near future.

Solar Sails

Solar sails, which are deployed in the airless environment of space after a conventional launch, are completely passive, using the pressure of sunlight to move as a sailboat uses wind to navigate on the ocean. Japan's space agency JAXA launched the world's first interplanetary solar sail mission, called Ikaros, in 2010. The 14-m-wide kite-shaped sail has solar cells embedded in the sail fabric to generate electricity. Utilizing a set of liquid crystal panels to change the surface reflectivity, the craft is able to change its orientation. Current flowing through the LCD panels increases reflection for Ikaaros to move forward, while turning the flow off reduces the sunlight pressure. JAXA also plans to send a 50-m solar sail with ion propulsion engines to the outer Solar System. This advanced sail hybrid would encounter Jupiter and some of the nearby Trojan asteroids.[5]

Other solar sail projects are in progress. The Planetary Society's LightSail-A spacecraft will go on a shakedown cruise to test sail deployment and engineering of the spacecraft itself. LightSail-B will demonstrate controlled solar sailing. The one-third-acre solar sail is scheduled to lift off in November 2014 on a SpaceX Falcon 9 booster.

Another Falcon 9 will launch Sunjammer, a large sail that will piggyback as a secondary payload with the National Oceanic and Atmospheric Administration's Deep Space Climate Observatory (DSCOVR) toward a gravitationally stable location called the Sun-Earth Lagrange Point 1, which lies about 1.5 million km from our planet. The largest sail ever flown, it will also be the lightest in weight. Sunjammer is designed as a precursor for such missions as early warning satellites for solar storms and systems for safely removing debris from low Earth orbit. NASA is combining efforts with engineers at the primary contractor, L'Garde Inc. of Tustin, Calif., to build Sunjammer.

5. The Trojan asteroids follow behind or in front of Jupiter, but within its orbit around the Sun.

In the 1970s, physicist Robert Forward detailed a different kind of light sail, one propelled by superlasers. A laser-driven solar sail could accrue higher velocity in shorter amounts of time than a solar-pressure driven one. In Forward's scenario, powerful gigawatt-level lasers would focus energy on the sail to propel it through interplanetary, or even interstellar, space. As the probe approaches its destination, the outer portion of the sail would separate from the central section. The outer sail would refocus Earth-bound lasers back onto the inner sail, slowing it down for capture at the planet of choice. Conventional braking systems could also be used for orbital insertion.

Another approach would be to use a maser to boost a solar sail composed of a mesh of wires with the same spacing as the wavelength of the maser's microwaves. Microwave radiation is somewhat easier to control than is visible light. Masers spread out more rapidly than optical lasers, owing to their longer wavelength, and so would not have as long an effective range. Masers could also be used to power a painted solar sail. Forward's hypothetical space sail probe could be coated with a layer of chemicals designed to evaporate when hit by the maser's radiation. This evaporation would, itself, provide thrust.

Plasma Rockets

More futuristic discussions have covered such areas as nuclear pulse thrust[6] and even antimatter. But today's engineers seem to be left with a choice: the fast acceleration and power of chemical rockets, or the high efficiency and painfully gradual acceleration of ion drive. However, there's a new kid on the block, an engine that promises high specific impulse, shortened travel times, and efficiency, and it's all the fault of a Costa Rican schoolboy who wanted, more than anything else, to be an astronaut.

Born in San Jose, Costa Rica, Chang-Diaz moved to the United States to finish his high school education, learning English in the process. From there, he excelled in school, earning a Ph.D. in Plasma Physics at MIT. He eventually did reach his vision of becoming an astronaut, and is the only NASA astronaut with dual citizenship. He has flown a record seven shuttle missions (including STS-75, which deployed the tether experiment satellite TSS-1R). But Franklin Chang Diaz has a passion for the future of human spaceflight, and he sees that future in plasma physics.

For two decades, Chang-Diaz has been developing the Variable Specific Impulse Magnetoplasma Rocket, or VASIMR, engine. VASIMR uses microwaves to heat propellant, turning it into a plasma. Says Chang-Diaz, "Our VASIMR engine is actually an electric rocket. It does not use chemistry, it does not use nuclear, it uses electricity. The electricity has to come from somewhere. So in our early applications of the VASIMR, we would use solar arrays that produce electricity. That electricity will be what we use to produce the plasma out of the working propellant."

6. For more on this, see George Dyson's book *Project Orion – The Atomic Spaceship 1957–1965* (Penguin).

VASIMR shares some similarities with other electric propulsion engines, Chang-Diaz explains.

> Dawn and Deep Space 1 are electric; they use xenon as the propellant and they use electricity to heat that xenon to make a plasma. Then, using electric fields, they extract an ion beam out of that xenon plasma. That's what the ion engine is. The same thing happens with the Hall thruster, that's the other type of electric thruster electric rocket in the field. The Hall thruster is a little bit more powerful than the ion engine. The ion engine is very low power, like a kilowatt. A kilowatt is about the power of a hair dryer.
>
> The Hall thruster has now gone to 4 or 5 kilowatts, so maybe five hairdryers. It's the same general approach. They take electricity, they heat the propellant, which is xenon gas, then they put it through the plasma, then that plasma is redirected in the form of an ion beam. And that's what happens in the Hall thruster also. The only difference is that the Hall thruster is four or five times more powerful.
>
> The difference between the VASIMR and other plasma drives is that VASIMR is able to process far more power. The VASIMR engine's power clocks in at about 200 kilowatts. 200 kilowatts is about the power of an SUV.... Now you're talking something more serious, more to the tune of something that can carry a fairly big payload in space.

Part of the problem with the lifetimes of ion engines is heat. When an engine runs for extended periods, hardware heats up, and working parts begin to fail. But the VASIMR uses a magnetic field to isolate the flow of the hot plasma, which Chang-Diaz likens to an invisible pipe. "What it does for you is that it gives you a pipe through which the plasma moves without touching the materials of which the rocket is made."

The magnetic field insulator serves as a liner that is impervious to the high temperatures that limit other types of ion propulsion. VASIMR is able to bring its plasma to millions of degrees, thus increasing its power.

The problem with a 200-kW engine is that it requires a 200-kw power supply. This is one of the limiting factors of VASIMR's technology. "You have to have a solar panel that is a lot more advanced, a lot bigger, a lot more powerful," Chang-Diaz says. How powerful a solar panel can you make? Well, it turns out that solar panels capable of producing 200 kW are now coming of age. For example, a solar panel the size of one of the ISS's solar panels is capable of producing roughly 100 kW today, and more efficient panels are in development.

However, ungainly solar panels lead some designers toward another option – nuclear power. VASIMR is not a nuclear propulsion engine, but the electricity to make it work might come more readily from a high-yield nuclear power plant, whose output is measured in megawatts, thousands of times that of solar power. And for humans, the kind of power generated by a nuclear source might be preferable on those long flights beyond the asteroids.

With those kinds of levels, says Chang-Diaz, "Now you're talking power levels like a 747 on takeoff. It's about 200 megawatts." Power levels like that would trim a human mission to Mars from 6 or 8 months to less than 2 months. The issue, he points out, is that "we don't have a 200-mW nuclear reactor. Not today. But that doesn't mean that we can't have one. It just means that we have to get to work on designing one and building one." He adds: "There's a lot of work that needs to be done in nuclear space power. That work is really being done at a very low level. People are very shy about this technology, at least in the United States." But the United States is not the only nuclear-capable nation. With a growing number of nuclear-capable nations joining the spacefaring community, nuclear space reactors may be developed by China, India, Russia or Europe. "The point is, we're not going to go very far in space if we don't develop these things."

Chang-Diaz sees innovation in technology as critical to accessing the outer Solar System, and he believes systems like VASIMR offer the kind of innovation we need. "I think the potential of the technology is tremendous. They call it game-changing. Disruptive. Those are the kinds of words that people use to describe the technology because in essence it can take as much power as you feed it. It can go to very high power levels. You will get performance that is extraordinary. I think what it will do is it will open up the entire Solar System to human exploration."

What is clear, analysts say, is that travel times to the realm of the giants must be decreased substantially. In addition to reducing exposure to the hazards of the space environment, reduction in flight times will mean reduction in every other consumable, from water to oxygen to food to fuel to power. Long-term travel will require sources of energy that are much greater or more efficient. As an analogy, Robert Zubrin points to the Spitfire, the RAF's star fighter airplane of World War II.

It had a 1,000 horsepower engine. Think about that for a minute. That's not terribly large by today's standards. But if you go back a hundred years from there, where everyone used horses, that's 1,000 horses! A thousand horses are pulling that airplane. There are planes today that have at least a hundred times the power of the Spitfire. So now it's 100,000 horses. That's as many horses as were in the army of Genghis Khan, pushing around one medium-sized airplane – like a 737 – and we've got thousands of them flying around every which way.

Former shuttle astronaut Alvin Drew comments

If you've got a good, powerful engine that's fairly fuel efficient, something like Franklin Chiang-Diaz's VASIMR engine, suddenly you're not doing this classic Keplerian coast to the outer planets. You're under power the entire way out there. Instead of getting a kick in the pants to get out to Jupiter and then getting another kick in the pants to get back, you are motoring out there to the halfway point. Then you turn around and you're putting on the brakes, decelerating into places like Jupiter and Saturn. You do the same thing coming back, so instead of spending years and years on these transits, you are spending months or even weeks coming back, and I think when that happens, then it becomes truly feasible to go explore the outer Solar System.

ARTIFICIAL GRAVITY

Flight surgeons consider one of the greatest barriers to long-term space travel to be the environment of microgravity. As soon as the human body is removed from gravitational stress, bone loss begins to occur.

No matter how much calcium is ingested into an astronaut's diet, bones continue to dissolve in the unrelenting microgravity of space.

Additionally, about half of the astronauts and cosmonauts suffer from space adaptation syndrome (SAS), a condition that includes severe nausea and disorientation. Gravity is integral to how the brain works out spatial orientation. The brain gets confused if it can't sense "down." Most astronauts can adapt to SAS over time, and can use patches or other medication to treat the nausea.

However, the physical effects of long-term weightlessness could be far more serious to the outer planet's traveler. In addition to skeletal weakening, some inner Solar System space travelers have been afflicted with vision problems. Cardiovascular systems weaken and muscles tend to atrophy. Astronauts in orbit spend literally hours each day exercising to combat these effects rather than using the time for mission-related activities.

Artificial gravity could solve all of these problems.

Much 1950s science fiction art is rife with illustrations of donut-shaped orbiting outposts and wheel-like space stations. Films depict giant rings in Earth orbit, spinning majestically to the tunes of *The Blue Danube* while their inhabitants sip coffee and congregate in international lounges with gently curving floors and ceilings.

The designs all assume that a spinning interior will cause objects to settle on the outside of the donut, which becomes the "floor" for the inhabitants. A spinning spacecraft can simulate gravity with centripetal force. Objects tend to move away from the center of the spin.

The approach has its problems, though. From an engineering standpoint, a spinning structure adds tremendous complexity to a mission design. The craft must be able to withstand the forces of a cartwheeling voyage, and must be able to carefully control its flight while spinning up for its cruise and then spinning down to a stationary mode upon its arrival.

One solution is to have only part of the spacecraft that spins, while the rest remains stationary. This scenario was successfully played out with the Galileo spacecraft at Jupiter (see Chap. 3), but a piloted mission would be on such a large physical scale that such issues as torque and interfaces might become problematic.

In 2011, NASA's Technology Applications Assessment Team developed a concept called the multi mission manned spacecraft. The goal of the spacecraft, named Nautilus-X, was to provide a platform for a six-person crew to engage in long-duration missions into deep space. Nautilus-X included a donut-shaped centrifuge area where the crew's living quarters would supply artificial gravity without affecting the rest of spacecraft operations.

Another technique for supplying artificial gravity is to spin a spacecraft on the end of a tether. This scenario was first attempted on the *Gemini 11* mission in September of 1966. After docking with an Agena upper stage, astronaut Dick Gordon left the spacecraft to attach a tether between the two vehicles. The spacecraft were then undocked, and *Gemini 11* moved to

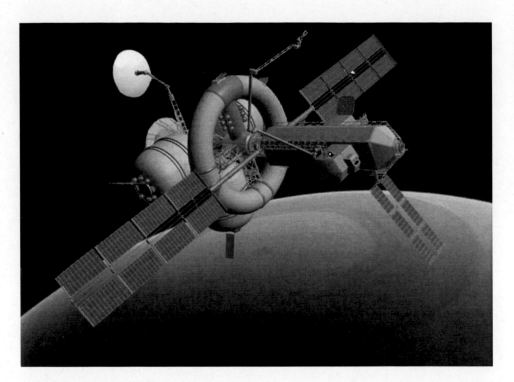

Fig. 9.6 The Nautilus-X, a proposed interplanetary ship using a toroidal spinning section to afford the crew gravity (Spacecraft art by Mark L. Holderman, NASA Technology Applications Assessment Team)

the end of the 36-m tether linking the two spacecraft. Spacecraft commander Pete Conrad initiated a slow rotation of the Gemini capsule around the center point between the two vehicles, trying to keep the tether taut and the spacecraft a constant distance away. Oscillations occurred initially, but damped out after about 20 min. The circular motion at the end of the tether imparted a slight artificial "gravitational acceleration" within *Gemini 11,* the first time such artificial gravity was demonstrated in space. After about 3 h the tether was released, and the spacecraft moved apart. Although the force of gravity was too minute for the crew to sense (roughly 0.00015 g), they observed objects moving toward the floor of the capsule.

Several other tether missions have been carried out. For example, the SEDS1 (Small Expendable Deployer System) spacecraft incorporated a 20-km-long tether counterbalanced by a spent Delta II upper stage. The space shuttle *Columbia* deployed the TSS 1R from the cargo bay, reeling out a tether for 19 km before a short circuit severed the cable. Both missions provided valuable insights into the application of tethers to future missions.

The tether technique for artificial gravity is a key component of several human space exploration scenarios, including Robert Zubrin's Mars Direct. In this mission concept, which details an inexpensive, incremental way to build an infrastructure of Earth-Mars expeditions, ships en route to Mars would tether themselves to the spent upper stage, using it as a counterweight for artificial gravity in transit.

Spinning a ship to create gravity may generate problems of its own. One concern involves the Coriolis effect. This phenomenon acts upon

Fig. 9.7 Right: An Agena upper stage floats at the end of a tether, seen out the window of Gemini 11. The experiment was the first using tethers in a human mission to simulate gravity in space. (Image courtesy of NASA/JSC.). Left: Mission scenarios such as Robert Zubrin's Mars Direct would use tethers to imitate gravity on the long transit journeys to and from Mars (Painting by Michael Carroll, courtesy of the Mars Society)

objects relative to the rotating frame of reference. Within a spinning environment, the apparent force acts at right angles to the actual rotation axis, curving the movement of objects in the opposite direction to the habitat's spin. For example, as an astronaut inside a rotating habitat moves away from the axis of rotation, he or she will feel a sensation of forces pushing the astronaut away from the direction of spin. These sensations confuse the operation of the inner ear and are disorienting, causing vertigo and nausea. Longer periods of rotation (longer spin axes) reduce this effect.

Several recent engineering studies have examined concepts trending in the other direction, using spin axes that are even shorter. The idea is that a short-radius centrifuge within a spacecraft can be used for limited periods, perhaps during periods of exercise. These small centrifuges spin rapidly, with the crew member rotating once every two seconds to create a force equivalent to Earth's gravity. Engineering studies supported by the National Space Biomedical Research Institute (NSBRI) in Houston, Texas, hope to determine whether or not short-radius centrifuge workouts produce the needed effects on bone, muscle and fluids in the body necessary to combat the detrimental side effects of microgravity.

Fig. 9.8 What a difference a little gravity makes! Salt and pepper, Earth style (behind), can be scattered from a shaker. In orbit, astronauts aboard the ISS must get their salt and pepper in a suspension of oil (foreground) (Photo by the author)

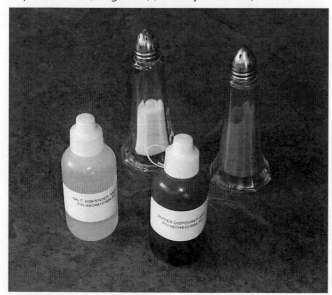

The most natural form of artificial gravity is simple acceleration. Says NASA's Alvin Drew:

Constant acceleration gives you a little gravity. That would help out tremendously. Think how hard it would be to make an omelet in microgravity. This is why all of our food is prepared for us on the ground and we simply reheat it in orbit. I could not prepare lox, eggs and onions as a Sunday morning breakfast once I'm out there, so it has to be done ahead. But what happens if I have even a tenth of a g on this spacecraft? Now, water stays in bowls so I can have standard plumbing like I have on Earth. Cooking is more of a challenge because oil is spattering in low gravity so that oil drop can hit you clean across the kitchen as opposed to standing over the skillet. Still, a little gravity goes a long way.

Some flight designers predict that the steady thrust of an advanced propulsion system will be the ultimate cure for weightlessness.

RADIATION SHIELDING

Two kinds of radiation will assault an explorer in transit to the outer Solar System. The first has a nearby source – the Sun. Our nearest star sends out barrages of high-energy particles, especially during solar flares, known as solar particle events (SPEs). These events can disrupt radio and television transmissions and can even disable satellites, but they also have the potential of delivering a lethal dose of radiation to an astronaut outside of Earth's protective magnetic field.

A second type of radiation comes from beyond the Solar System, out in deep space. It is cosmic radiation, and its specific sources are unknown but may include such overly energetic places as supernovae and galactic cores. Cosmic rays pose a host of greater health risks in the long term, including cancer, liver disease, damage to DNA and the central nervous system, cataracts and other disorders.

Fortunately, crews can be shielded from solar radiation. Although heavy, some kinds of metal can provide protection from solar flares. Several feet of water or pure hydrogen will form an effective barrier from even the most deadly of SPEs. Ships can be designed with "storm shelters," small chambers where crews can retreat during such events. These shelters can be surrounded by the ship's fuel, especially if it is hydrogen, or the walls can be filled with the crew's water supply. As the water is used up during cruise, secondary walls can be filled with waste to retain the sheltering effect.

Cosmic radiation is another matter. Its dense particles, stripped of all electrons, sail easily through solid metal and even the cores of planets. The biome of Earth is shielded by the planet's magnetosphere, the protective bubble surrounding the globe that shunts solar radiation away from the surface. This field also helps to shield us from cosmic rays.

Since the 1960s, engineers have wondered if a similar solution could be found for a spacecraft. Could a crew be protected by an artificial magnetosphere surrounding their ship? A recent study done by physicists from the UK, Portugal and Sweden demonstrated that the technology could be made compact enough, and therefore cheap enough, to protect an inhabited spacecraft in deep space. Like a planetary magnetosphere, these local fields would generate a charge in the space around the craft, deflecting the deadly particles away from the spacecraft. Using an electromagnetic probe as a mock spacecraft, the team realized that their magneto-ship deflected plasma away, clearing the space immediately around it. They estimated that a spacecraft could be protected within a magnetic bubble roughly 100–200 m across, using a system that could be readily carried into space.

Another recent study, called SR2S, is being carried out by CERN and several French, Italian and international European partners. SR2S hopes to develop several key enabling technologies to shield crews from radiation. Their goal is to generate protective magnetic fields around their craft using superconducting magnets.

Researchers at Johnson Space Center have been experimenting with high temperature superconducting structures to generate protective fields. Their studies suggest that, "a combined system of active and passive radiation shielding constitutes the most promising solution to this issue." JSC engineers envision large, ultralight, expandable coils to "reduce radiation exposure of humans in the spacecraft habitat to acceptable levels over longer duration missions."

Franklin Chang-Diaz's VASIMR engine generates its own magnetic field. The field might also be modified to serve as crew protection, he says. "It's a very interesting byproduct of the architecture of the engine. You have a very strong magnetic field. Yes, you could consider developing a magnetic shield around your ship that would shield crew members from some of the radiation."

The technology VASIMR uses to create its magnetic fields is superconductivity, the technology of superconducting magnets. This technology is another one that has come of age, he says. "There are some really amazing superconductors now that can carry very high levels of electric current that produce very strong magnetic fields. These are very lightweight conductors. They operate at temperatures that are far above those of the early superconductors, which had to be chilled down to nearly absolute zero. You don't need that anymore."

Outer planet explorers will contend with yet another source of radiation once they arrive at their destination, the energy that comes from the cores of the planets themselves. The magnetospheres of the gas and ice giants are relatively benign, with the exception of Jupiter's. In all cases, these energetic fields can easily be avoided with distance from the planet itself. In the case of the Galilean satellites, two important targets, Europa and Io, lie deep within Jupiter's radiation zone. Visitors might be able to find sheltered regions (see Chap. 10), but most research may well be carried out using telerobotics, controlled from a nearby base on Ganymede or Callisto.

One other implement that future astronauts may find in their anti-radiation toolbox is not metallic or electromagnetic, but rather prescription. Drugs may provide an answer. Radiation's destructive power is so insidious because of scale. It damages genetic material inside of our cells, and it destroys brain cells. But a class of drug known as radio protectors has been used successful in patients who have been exposed to radioactive environments, and several drugs even show promise when administered ahead of time. Genistein is a type of antioxidant. Not only is it non-toxic, it naturally occurs in some foods, and can be administered long-term without side effects.

Synthetic triterpenoids, compounds derived from plant mosses, also have powerful antioxidant effects. Recent laboratory tests[7] show that the drug protected mice that were given the drug before radiation exposure. Filgrastim has been used on patients who have received high doses of radiation, and whose white blood count is depressed. It has not yet been approved specifically for radiation treatment, but studies continue. Yet another drug, Amifostine, is regularly used to reduce the side effects of radiation treatments in cancer patients, and is on NASA's short list for possible future use in long-duration space flight.

HUMAN HABITAT SIMULATIONS

But aside from radiation, how can astronauts be tested for the effects of long-distance space voyages? And what issues will face crews living long-term in a glorified "can" on the surface of a distant world? Flight planners are gathering data from a multitude of sources, but it hasn't always been easy. Researchers have sometimes likened duty aboard submarines or naval ships to long-term space travel.

In the 1970s, the Navy Submarine Medical Research Laboratory in New London, Connecticut, began collecting medical data on submariners who volunteered to be medically followed over several years. Similarly, a 1,000-person aviator study was carried out by the Naval Aerospace Medical Research Laboratory in Pensacola, Florida. In both cases, the subjects were more closely scrutinized medically and psychologically because of their unique career path, not in the context of any extreme environmental conditions.[8] But with a built-in infrastructure of medical tracking, submariners and others in the armed services could provide valuable sources for future studies applied to long-duration space travel.

Scientists have been conducting mission simulations since the Cold War era. The Mars 500 mission began in Moscow in 2007, but it had its roots in the heady years of the Man-on-the-Moon programs. Back in November of 1967, the Russian Academy of Science's Institute of Biomedical Problems (IBMP) stuffed three volunteers into an isolation chamber affectionately referred to as "the barrel" for a year. It was the first of a series of expanding experiments designed to test humans in closed-loop environments, where water and air are recycled. By 1970, the facility had grown to a three-module simulator called the Marsolet, or Mars flyer, which included a greenhouse. Future tests raised the bar considerably. The endurance experiment included Russian, ESA and Chinese researchers and subjects. Three different crews rotated through a complex of modules that simulated a living habitat module, medical facility and utility module, all

7. Tests were conducted at the University of Texas Southwestern Medical Center in Dallas, with results presented at the American Society for Gravitational and Space Research in Orlando, Florida, November 2013.

8. Personal communication with Dr. Steven Linnville, Deputy Director of Research at the Navy Medicine Operational Training Center, Pensacola, Florida.

Fig. 9.9 The exterior of the Mars 500 habitats. The Mars 500 facility housed a crew for 17 months. The airlock and Mars environment are to the left (Image courtesy of Mars 500 project, © IBMP/Oleg Voloshin)

assembled as the primary spacecraft. A Mars Transfer Vehicle, connected to a Martian surface environment area, was only used by three crew members during the 30-day "Mars exploration period." They spent 1 week in the simulated Martian environment, carrying out three Mars EVAs (extravehicular activities, or Mars walks) before "returning" to the mother ship complex.

The final mission ran for a total of 520 days in the enclosed faux space complex. The six-person crew included three Russian members. ESA supplied a French participant and an Italian-Columbian subject. China rounded out the crew with the sixth volunteer. The crew climbed aboard on June 3, 2010, with a realistically limited supply of resources. The modules carried data and control systems, communications equipment, air and water supplies and partial recycling systems. Flight controllers built in an ever-increasing time lag to simulate the light-distance from Earth to the Mars-bound craft. Contact with the outside world was realistically limited as well.

The majority of the crew suffered from lack of sleep, and some became lethargic to the point that they no longer did prescribed exercise or work tasks. A 2013 report[9] to the National Academy of Sciences revealed, "crew sedentariness increased across the mission as evident in decreased waking movement (i.e., hypokinesis) and increased sleep and rest times." Members also experienced a disruption of sleep-wake cycles, "suggesting inadequate circadian entrainment." The experimenters concluded that future crews will need to travel in spacecraft environments that mimic terrestrial physical characteristics (for example, 24-h day/night cycles and regular meal times).

Some engineers would like to see the next step as a full simulation involving a two-person crew at the ISS. Under this scenario, two astronauts would travel to the ISS, spend 9 months there as a simulated cruise to Mars, then return to Earth for 1 month. After the brief stay on Earth, they would again ascend to the ISS for another 9 months before returning to Earth at end of the mission, giving flight engineers, psychologists and medical staff the opportunity to carefully evaluate what such a flight scenario would do to the human body and mind.

Shuttle veteran Alvin Drew says that familiarity may, in the long run, profoundly affect mission success. "You're spending a lot of your cognitive load trying to adapt to a new reality. A lot of the things you take for granted,

9. Mars 520-d mission simulation reveals protracted crew hypokinesis and alterations of sleep duration and timing. Mathias Basner et al., PNAS, accepted November 27, 2012. See also http://www.imbp.ru/Mars500/Mars500-e.html.

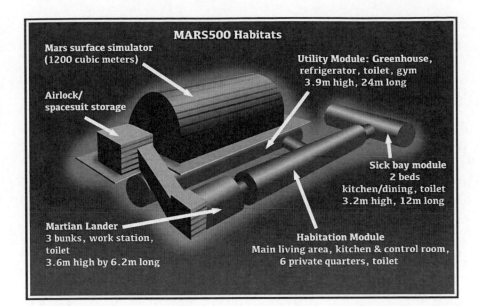

MARS500 Habitats

Mars surface simulator
(1200 cubic meters)

Airlock/
spacesuit storage

Utility Module: Greenhouse,
refrigerator, toilet, gym
3.9m high, 24m long

Sick bay module
2 beds
kitchen/dining, toilet
3.2m high, 12m long

Martian Lander
3 bunks, work station,
toilet
3.6m high by 6.2m long

Habitation Module
Main living area, kitchen & control room,
6 private quarters, toilet

Fig. 9.10 Modules in the Mars 500 facility included living quarters and a Mars vehicle and Mars "environment" pod (Author art based on ESA diagram)

like how you balance and move, suddenly become part of your conscious brain. We all have this thing we call the 'space stupids.' We always assume our IQ is cut in half when we first go into space." Things that seem intuitively obvious on the ground require practice ahead of time, because when astronauts get into space, simple things can just stymie them. Drew attributes the phenomenon to the fact that "I'm trying to adjust to this new world where things that should work normally don't. The way you store things isn't the way you store things on the ground, the way you go get water or prepare your meals is altogether different; now your brain has to go and spend a lot of its cerebral capacity doing things that it's usually pushed down into lower brain functions, and now your upper brain functions are having to deal with it." In Drew's experience, crews adapt over time by creating surroundings similar to daily living environments already familiar to astronauts.

Fig. 9.11 A Christmas tree, made primarily of the spines that held food cans during launch, brought a touch of home to the astronauts on board the Skylab space station. Video image from Skylab 4 (Image courtesy of NASA/JSC, Ed Gibson)

As an example, Drew points to driving a car. Few seasoned drivers devote much attention to the mundane tasks of turning the key, working the brakes, adjusting the mirrors or checking the blind spots. These things become automatic. "Now imagine," says Drew, "if I put all that stuff over in the passenger side of the car, how much more difficult your task becomes. We do that a lot in spacecraft, making things different than they need to be."

At roughly the same time the Mars 500 tests were ensuing, NASA ran a series of experiments, still on-going, called Deep Space Habitat. NASA's Alvin Drew was a participant:

The Deep Space Habitat study was looking at the crew modules and crew-inhabited portions of an interplanetary spacecraft, for things that would go to Mars or beyond. We were looking at missions that would be 500 days or longer. It was basically a sandbox where you would bring your technologies for long-duration space missions. We'd put them in a place where everything was integrated. For example, if you had a concept for lighting, would that work well with our power scheme or with the data handling? So everybody comes together and brings out their stuff, and my objective was to break their things for them and tell them what was wrong with it so they could fix it up.

Much of the on-board work involved a sort of rock triage from an asteroid, deciding which samples were most important to bring back directly to Earth with the crew and which ones could stay in orbit for later pickup. As in the Mars 500 study, time delays were built in to communications, simulating ever-increasing distance between craft and home. Drew's crew was tasked with simulating the return-leg of a voyage to an asteroid, so the team was subjected to delays of 5 min, 50 and 10 s as their mock ship closed the distance to Earth.

The issue of time delay with great distance has long been a challenge in outer Solar System exploration.[10] Alvin Drew suspected that the longer the delay was, the more difficult it would be. But what the crew found out was surprising. "It turns out that there is a nadir where it's most difficult. For me it was the one-minute time delay. Beyond that, you have a kind of shift where you become more autonomous as a crew. You don't feel compelled to tell everything to Mission Control because it's too much of a bother. Inside that, you will make that attempt and it gets to be kind of tough. You're in the middle of a transition period between where you're autonomous and where Mission Control is running all the detailed aspects of the mission."

10. *Voyager 2* missed several precious moon encounters at Saturn because of a jammed scan platform, which prevented accurate pointing of the instruments (see Chap. 3). By the time flight controllers knew there was a problem, Voyager was exiting the Saturnian system.

Alvin's crew found texting to be the most efficient way of communicating in those long time delays. "Let's say you're reading off some long numerical sequence and they say, 'Repeat again.' If you've got a five-minute delay that conversation could stretch out into twenty minutes." The Deep Space Hab team found it was better to use an instant messaging application. There is no easy way to have a two-way conversation with the long delays involved in outer system exploration. Many communications are laden with complex, technical data. For example, a typical message might say, *If you look on line 21, procedure 203 decimal 9, the data value there is 2.1 microvolts.* If any of that transmission gets garbled, or just missed, the easiest thing to do is have a written record.

Astronauts also looked into using one of several speech recognition programs. Often, crew members don't have the time to key in data, or they are physically unable to do so due to wearing a spacesuit or using hands in a glove box or control console. In those cases, astronauts would simply speak into a microphone, check the message on the screen, and give a voice command like, "Yeah, like that. Send it." For the most part, it worked really well. But not always, says Drew. "There was an instance where we actually had to disable it. We had an X-ray spectrometer that we used to figure out what the materials were in different rock samples. It would send out X-rays and that was a hazard. If we said the wrong word or phrase in passing, like 'turn on,' this thing would fire itself up and start shooting X-rays. We decided it was best to manually operate the system than have it misinterpreting things we were saying."

OTHER SIMULATION PROGRAMS
FOR DEEP-SPACE TRAVEL

The Canadian Arctic will play host to another ambitious simulation, Mars Arctic 365, at the Mars Society's Flashline Research Station. Just 900 miles from the North Pole, the facility was completed on Canada's Devon Island in the summer of 2000. The cylindrical Mars Habitat stands 7.7 m tall. The interior diameter is 8.3 m across. The first floor contains two airlocks, a shower and a toilet, spacesuit storage and a laboratory/work area. The second level houses six crew rooms with bunks, a common area, a kitchen with a gas stove, refrigerator, sink and microwave oven. Above this level, a ladder leads to a loft that can be used for a seventh crewmember or storage.

Devon Island is an ideal setting for a long-term space simulation. Its Arctic terrain holds many geologic analogs to Martian landforms, such as mass wasting from underground ice, patterned ground in permafrost regions and even an impact crater called Haughton Crater.

The Mars Arctic 365 program will subject researchers to the most extensive simulation to date of a multi-person, full-scale mission in isolated, harsh conditions. Areas of research will include geology, microbiology, human physiology and psychology, nutrition-food investigations and technology experiments involving remote rovers and field stations. As in all previous simulations, if any work is to be done outside, it must be done in a spacesuit. Travel is by ATV.

Mars Arctic 365 differs radically from simulations like the Russian Mars 500, says Mars Society leader Robert Zubrin.

> For Mars 500, they had people in a room in Moscow. They did not have to get any real science done. They were not faced with the challenges of fieldwork. They were not faced with the challenges of balky equipment failing under field conditions, not faced with the need to fix generators in the cold. They were not facing physical risk, not truly isolated. Being isolated is not being in a room and not being able to go out and socialize with people. Being isolated is being separated from this global network of production and distribution that is our life support system on this planet. In other words, right now, you have access to the products and skills of 7 billion people – the whole world market. You have access, for a price, to the medical expertise of early 21st century civilization. If anyone on Mars One had a case of appendicitis, they could have just left the place and been taken care of. There was never any issue of their ability to access the full life support and protection of civilization. On Devon Island, we don't have that. It's a very tenuous connection. If you have an injury, you may be able to get a Twin Otter in that day, or maybe not for two weeks. We're not on Mars, we're on Devon Island. But there is real risk. There is real danger.[11]

Still other ongoing tests of humans in space are being carried out on the International Space Station. Typically, crews rotate through the outpost every 6 months, which is about the time it takes for a Mars cruise. But in 2015, three team members will begin a year-long stint in the microgravity environment in orbit.[12]

11. For more on the project, see http://ma365.marssociety.org.

12. The ISS has the capacity to host six people at a time, limited by the capacity of spacecraft standing by to evacuate the total crew at any given time.

Fig. 9.13 The Flashline Mars Arctic Research Station on Devon Island in the Canadian Arctic. The three-story module will host a crew for a 365-day simulation of a Mars mission (© Joe Palaia/Mars Society. Used with permission)

Scott Kelly will be the American crewmember on board for a 1-year-long mission. He's already had a 6-month stay in orbit, so researchers will have some side-by-side data. But Kelly offers another advantage. He has an identical twin brother who has flown two shuttle missions. Mark Kelly has never flown a long-duration flight. The comparison promises to supply good insights into microgravity effects on health and what role is played by genetics and environment.

Two cosmonauts will stay up for the year as well. There is nothing unprecedented about this in Russia.[13] The record for long-duration spaceflight is held by cosmonaut Valeri Polyakov, who was in space for 438 days on the Mir space station.

Researchers hope to gain valuable insights as they compare results from the ISS year-long voyage with the simulated Mars 500 and Mars Arctic 365 surveys.

FOOD PRODUCTION

One element of familiarity for astronauts – whether on a week-long orbital flight or a years-long outer planets voyage – is found at the dinner table. Space psychologists have known for some time that the types, diversity and quality of food affects crew morale and sharpness.

They say you can't take it with you. To many spaceflight designers, this adage applies to food supply. On a Mars-bound ship or a mission of a few months, prepackaged food is the modus operandi. But on longer voyages, food production may become an important part of crew health, both physical and mental. Experiments have been done aboard shuttles, Mir and the ISS. Astronauts and cosmonauts have grown radishes, lettuce and other greens in modest quantities. Crews have clearly established the feasibility of growing vegetables in microgravity. Experience has shown that the fresh quality of food, even in conjunction with packaged meals, is psychologically important. Alvin Drew comments. "I know the Progress [supply] capsules fly up a lot of fresh onions and apples. That's usually a highlight for those guys when those supply ships arrive and they can smell all those fresh vegetables, aromas emanating out of that capsule when they crack the hatch open."

Not all of the research is being done in space. Six Cornell scientists conducted a 118-day cosmic culinary mission at an altitude of 2,400 m on the slopes of Hawai'i's Mauna Loa. Called HI-SEAS (for Hawai'i Space Exploration Analog and Simulation), the 4-month trip was tailored specifically to determine what best to feed crews during long space journeys. The team ate a combination of prepackaged foods and foods that they prepared on site using shelf-stable ingredients. Members rated meals in detail, and kept copious records of their moods, body mass and general health. Of importance to the entire group were the use of spices, herbs and hot sauce. Comfort foods such as Nutella, peanut butter and margarine

13. Russians also hold the next highest space endurance records: 380 days by Sergei Avdeyev aboard Mir from August 1998 to August of 1999; 365 days by Vladimir Titov and Musa Manarov aboard Mir from December 1987 to December 1988; 327 days by Yuri Romanenko on Mir, from February 1987 to December 1987. American Sunita Williams holds the record for the longest duration flight by a woman. Her 195-day mission aboard the ISS began on December 10, 2006, and ended with the STS-117 shuttle landing on June 22, 2007.

also rounded out the top of the list. Foods rich in fiber will be an issue, as stored foods are usually highly processed and lacking in fiber. Wheat and rye crackers, nuts, and dried fruits played an important part in foods that last well over time.

Says, Alvin Drew, of food cravings:

> My personal experience was that on my first mission my tastes shifted. I had heard that this could happen. The first one was string beans. I wouldn't go five steps out of my way to get an extra helping of string beans at home. I got to orbit on STS 118, my first mission, and I just had this craving for the dehydrated string beans and mushrooms. I was trading away other foods I like to get everybody else's string beans. I just never had a jones for string beans all of my life. It surprised me to have such a fascination with string beans all of a sudden. I got home and everything was back to normal. I don't like string beans any more than I did before I left. On my second mission, I didn't have any of those weird anomalies, no cravings (aside from barbecue, but I always crave barbecue).

CLOSED-LOOP SYSTEMS

One important benefit of growing fresh food within the confines of a cruising spacecraft may be the production of oxygen and the recycling of water. As missions become longer and longer, the importance of "closed loop," or recycling systems, becomes more critical. A closed-loop system attempts to recycle critical elements like oxygen and water, converting waste elements back into usable resources. In the 1960s, Soviet scientists at the Biophysics Institute developed an experimental closed-loop life support habitat called BIOS-3. The underground center used primarily algae and plants. By 1972, two-member crews were spending 5 months in the 315 cubic m base, growing their own plants for food and oxygen. The loop was so efficient that only 20 % of the food needed to be brought in from outside.

In 2005, ESA added support to the ongoing BIOS-3 station, but the space agency has plans of its own. ESA has undertaken an experiment, called the MELiSSA project (Micro-Ecological Life Support System Alternative), an ecosystem based on microorganisms and higher plants to create a closed-loop biome for astronauts. Waste products and air pollutants from the crew and from various spacecraft systems are processed using the natural metabolism of plants. Those plants provide food and purify water and oxygen for air recycling.

MELiSSA consists of four compartments. A liquefying compartment collects mission waste as well as the non-edible detritus from the plant compartment. This heated chamber uses anaerobic bacteria to transform waste into forms usable by other compartments. The second compartment breaks down fatty acids, while the third one transforms waste to nitrates for use by higher plant forms. The fourth chamber uses algae and higher plants to produce oxygen and food.

MELiSSA is more advanced than other recycling systems used on Mir or the ISS. While the orbital experiments have purified water and recycled exhaled carbon dioxide, neither attempted to recycle organic waste for food production. ESA scientists continue to work toward this goal.

Experiments with microorganism oxygen production are currently being carried out on the ISS. Similar work has been done on the Soviet/Russian MIR and on the earlier Salyut space stations.

AUTONOMY

In Stanley Kubrick's classic film *2001: A Space Odyssey,* the crew of the Jupiter-bound *Discovery* spacecraft are dogged by the homicidal efforts of the ship's psychotic computer, the HAL 9000. But behind author Arthur C. Clarke's computer lies an important question. How much autonomy can we have in an advanced piloted mission? How much do we need?

Shuttle astronaut Alvin asks, "Can we have artificial intelligence to help us run our spacecraft? I like to think of ourselves as astronauts as being very smart, but even a crew of six astronauts is nowhere near as smart as having six dozen people at Mission Control with strip charts and computers and all that training. How much of that can I put on board the spacecraft – our own version of HAL? It could tell me, for example, 'Based on what I'm seeing for signatures, you're about to lose one of your generators on your Bus B. The best way to fix that is to shut this down over here and, oh, by the way, this is going to affect some of your experiments and timelines.' It can be a sort of associate or virtual Mission Control that's along with you. Especially if you're out at Saturn, where you are one light hour away from Earth, having Mission Control work those kind of details is simply not tenable."

The radio delay to Earth is not the only need for autonomy. In a reconnaissance mission involving a giant world with dozens of moons, multitasking will be the word of the day, and astronauts will need help. Remote-sensing instruments can be tasked to assist. JPL's Ashley Davies worked with the Earth Observing One satellite system to search for active volcanoes on Earth. Davies believes autonomy is the best approach for many robotic situations. "This is often the best thing to do: put autonomy on board your spacecraft and rovers so that they can react in real time to the detection of dynamic events. On board the satellite is an autonomous monitoring system designed to process hyperspectral data on board the spacecraft. It looks for specific spectral signatures of interesting dynamic and, in many cases, transient events, and if it finds them, it retasks the spacecraft to obtain more data."

This real time capability trumps sifting through data and writing a computer program to find a volcano that may have already finished erupting, he says. "It's flying in Earth orbit and doing rather well for itself. We've looked not only at volcanoes but at fires, other thermal events, the formation and breakup of ice sheets, flooding events. We're looking ahead to future missions where we can put the same kind of software on other spacecraft, where data can be processed to return." He adds, "This overcomes a major problem with communications. You can always collect more data than you can send back, so surely it's better to have some algorithms that run through the data that you can't send back looking for specific things or some outliers or things that you just can't explain." The spacecraft software either flags that part of the dataset or you send back a fraction of the huge dataset as a sample. Instead of waiting 2 or 3 weeks for an observation, researchers can obtain it in 90 min. In the meantime, the spacecraft has already tasked itself to get more data.

"You can imagine how this would work on a flyby past Io," Davies says:

> As you're coming into Io, the software is looking at where there's interesting stuff taking place. If you had a very flexible operational plan, you could then target specific areas for when you're closer in to Io. If you see something that's incredibly bright, you can reset your gain state on your instrument or default to a different sequence of observations so that you take different exposures at different times to avoid saturating the data. This, actually, is a major problem on Io, because if you're coming in and you see a really bright eruption, and you want to get unsaturated data close up, you're sacrificing all the information from everywhere else on Io – it's going to be terribly underexposed. If you're looking at the background of Io itself, you're going to be missing all your data at the large eruption sites. This is a way of dealing with that. It shows how you can squeeze more science return out of missions out of use of these autonomous systems.

In concert with the operations of talented crews, autonomous systems on spacecraft will enhance the productivity and safety of human voyages to the outer worlds. The nature of these distant worlds is foundationally dissimilar from the nature of places that astronauts have explored already. The nature of the spacecraft, flight profiles and strategies that will take us to the giants must change to meet those differences.

As our missions to the giants will be of a different nature from the lunar and Mars expeditions on the drawing tables today, so, too, our civilization will be a different one. We've gone from a civilization dependent on paleofuels (oil, coal) to one learning how to operate more cleanly and efficiently, graduating to biofuels, geothermal, hydroelectric and solar energies. If done responsibly, these sources are all sustainable. And the promise of fusion knocks at our door. In the discussion of settling the outer Solar System, some would consider the exercise one in fiction or, at least, hubris. But others see it as a natural outgrowth of what has come before. With its inventories of rare earths, potential energy sources like helium 3 and other unknown resources, the outer Solar System may be a key element in tomorrow's resource landscape for humankind. But the worlds beyond Mars beckon in other ways. Humans are destined to explore, to expand their areas of influence, to satiate their curiosity for the betterment of humankind and planet Earth, and to wonder at the world – or worlds – around them. Beyond the mundane problems of nuts and bolts, breadfruit and bamboo, what awaits us out there?

Fig. 10.1 A cruise ship off the "coast" of Titan. Future travelers may tour the satellites of Saturn in much the same way that cruise liners of today tour the islands of Greece or the Caribbean (Painting © Michael Carroll)

Chapter 10

Frolicking in the Outer Darkness: The Cultural Side of Living Among the Giants

What compelled the first settlers to leave a comfortable, prosperous and safe life to brave treacherous seas, rugged mountain ranges or perilous deserts? What compels us to go forward into the cosmos? Aerospace engineer and visionary Robert Zubrin believes that exploration is part of a built-in feature of the human species. "The innate desire to go where no one has gone; to build where no one has built; to do what no one has done … to see and do and go where no one has gone. That is the fundamental drive that brought us out of Africa. We only live in other parts of the world because we developed technology."

For some, the draw of celebrity, of being the first heroic explorer, may be a primary driver. For others, the reasons may be of a more spiritual or psychological nature. Fame and notoriety alone, however, cannot explain the risks and investment people are willing to make. Nor can simple fortune. Elon Musk, founder of PayPal and of the aerospace company SpaceX, has taken risks to establish a company that enables space exploration. He was not motivated by finances, Musk told a graduating class at Caltech. "Going from PayPal, I thought: 'Well, what are some of the other problems that are likely to most affect the future of humanity?' Not from the perspective, 'What's the best way to make money?'"

Joe Palaia, civil engineer and Business Development Manager at Pioneer Astronautics, adds that, "Maybe the people on the tip of the spear will do it for the immortality above all else, but other people who are involved will see other reasons for doing it. Profit motive will be a big one for some. If you think of Levi Strauss, he didn't necessarily join the Gold Rush because he wanted to be immortal. He wanted to get rich by selling jeans to miners! And there's also another appeal. Instead of being one of many, you can be one of a few. On a frontier, there are more opportunities."

Explorers Lewis and Clark surveyed the western United States at the bidding – and funding – of the U. S. government. The opportunities of the Pacific Northwest were not yet apparent to those in the East. But a profit was there for the making, there were practical and viable ways of carrying out travel and trade, and the risk-to-reward ratio could be made right. Once the Corps of Discovery had surveyed the Pacific Northwest, the transcontinental fur trade began. At a certain point, when an infrastructure had been set down, settlement could begin in earnest.

The cosmic realm of the giants, however, is of a new scale, says space artist Jon Lomberg. "Talking about settling the outer Solar System now is like the pilgrims trying to worry about how they were going to cross the Rocky Mountains. What is going to motivate something to go out there? Will that something be machine or human? I think it will depend on so many factors that we are completely ignorant of. The motivation and the means are at a different level of discussion than making a colony on the Moon or a colony on Mars."

M. Carroll, *Living Among Giants: Exploring and Settling the Outer Solar System*,
DOI 10.1007/978-3-319-10674-8_10, © Springer International Publishing Switzerland 2015

Fig. 10.2 Lewis and Clark meet a tribe of Chinookan Native Americans on the Lower Columbia in October of 1805 (envisioned here in a painting done a century later by Charles M. Russell). Lewis and Clark's Corps of Discovery opened the way for the fur trade and eventual settlement by the Europeans. (public domain)

Nevertheless, as humankind continues to traverse the Earth-Mars void, technologies will advance, making the crossings easier and more cost-efficient. Robert Zubrin says that we need look only to history for evidence of what is to come. "Columbus crossed the Atlantic in very primitive ships, but 50 years later there were caravels, and then there were clipper ships and then steamships and Boeing 747 s. The grandchildren of the first Mars colonists will hardly believe the stories of what the immigrant experience was like in the early days," he says. He adds: "The same capabilities that make going to Mars easy will make going to the outer Solar System possible, and when we start settling the outer Solar System, the same technologies that make going to the outer Solar System easy will make interstellar voyages possible."

Engineer Joseph Palaia sees this incremental approach as key to the future. "It's going to take mass to settle these places in the outer Solar System. If we have an industrialized society on the surface of Mars, that's a smaller gravity well than lifting all that stuff up from the surface of Earth. It could very well be that we see a progression of industrial capability out into space. We go to Mars and establish ourselves there before we go farther."

But before that infrastructure is in place, humans must blaze the first trails.

Fig. 10.3 *Mars may play a key role in the industry required to develop infrastructure among the giant planets. Volatiles at the poles offer many processing options (Painting © Michael Carroll)*

THE FIRST LANDINGS

Approaching a giant planet is like approaching a complex miniature Solar System. Many moons in varied orbits, some moving opposite to others, and some orbiting in planes other than the plane of the planet's equator and rings, all make for a challenge to any navigator. But flight engineers have had practice with multiple planetary flybys, and with orbiters such as Galileo/Jupiter and – most recently – Cassini/Saturn. Cassini's Carolyn Porco emphasizes that, "Cassini has been a mission of unprecedented magnitude. We have conducted more close flybys with Cassini – threading the needle, making sure your speed vectors are just right and then you clean up afterwards, precision maneuvers – than have ever been done in the entire space program – well over a hundred."

With all that experience, Porco suggests that getting into orbit or even on to the surfaces of various satellites would be old hat. But with people, a spacecraft must carry far more mass to keep crews alive and healthy. In many cases, the satellites are so small and of such low density that landings and launches require very small amounts of energy, even for a massive piloted spacecraft. At Enceladus, for example, explorers will not have to worry about heavy radiation shielding. In fact, they will scarcely need to worry about getting off of Enceladus at all, with its low gravity. The main concern will be getting out of the primary planet's orbit, a case common to all the outer giants.

One early strategy of outer planet exploration may be to establish a beachhead on the smallest moons. These mini worlds have the advantage of proximity to the giant planets and their large satellites, while having the low gravity from which to come and go. At Jupiter, in particular, the small moons Elara, Himalia and Leda orbit outside of the deadly radiation belts, but near enough to operate telerobotics on the major Galilean satellites. Leda is closest and smallest.

At Saturn, Pandora, Prometheus, Janus and Epimetheus, would provide close bases from which to research the major moons, but they would also afford spectacular edge-on studies of the ring system and its complex

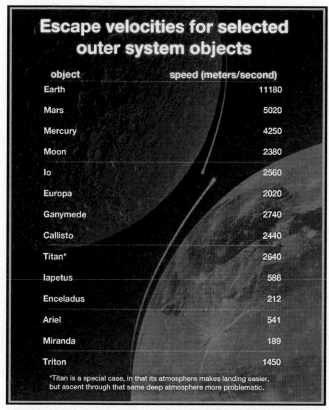

Escape velocities for selected outer system objects

object	speed (meters/second)
Earth	11180
Mars	5020
Mercury	4250
Moon	2380
Io	2560
Europa	2020
Ganymede	2740
Callisto	2440
Titan*	2640
Iapetus	586
Enceladus	212
Ariel	541
Miranda	189
Triton	1450

*Titan is a special case, in that its atmosphere makes landing easier, but ascent through that same deep atmosphere more problematic.

Fig. 10.4

wave dynamics. The smallest moons of Uranus would provide no real advantage, as the mid-sized Uranian satellites (Miranda, Ariel, Umbriel, Titania and Oberon) have low gravity anyway. At Neptune, some of the smaller moons might be tempting targets, simply because Triton is a difficult place to land. Triton orbits in a retrograde direction, opposite to the natural orbital motion of moons in the system or approaching spacecraft. Because of this, Triton's landing speeds are high. Its gravity is also substantial. The moon's orbital movement actually gets any departing spacecraft started in the wrong direction, so departures take far more energy than ones from the smaller pro-grade moons. Among these targets will be Proteus, Larissa, Galatea and Despina. At a distance of up to 10 million km, Nereid's oddball orbit may range too far away from the Neptune system to make it a practical observation or telerobotic site.

Wheaton College's Geoffrey Collins is part of the team that recently assembled the most detailed geologic map to date of Jupiter's largest moon Ganymede. "One of the primary things we were thinking about [in doing the map] was future landing sites. I think there are places of interest. I personally would rather reconnoiter the entire surface before spending a lot of money to land something on it. We don't even have a complete global map of the surface. There are still large areas of the map where we have to draw in mermaids and dragons. There are definitely some fuzzy bits left on it."

A greater percentage of Europa has been mapped in detail than has Ganymede. Callisto has been imaged in detail on less of its surface, and Io has a percentage of high resolution somewhere in between. The best mapped of Saturn's satellites is Enceladus. Cassini has a total of 22 scheduled flybys of the geyser-riddled moon. Moderately to finely detailed maps are now also available for Iapetus, Dione, Rhea, Mimas and Tethys.

Detailed maps of Titan are being assembled with nearly each pass of the Cassini spacecraft, but radar provides narrow strips of the landscape, known as noodles, that must be painstakingly pieced together. At Uranus, all the moons during the Voyager encounters – which provided our only detailed views to date – had one hemisphere permanently in shadow (see Chap. 8). Maps of the five main moons vary greatly in resolution, with Miranda and Ariel having the highest resolution. At Neptune, the only moon imaged in detail was Triton, and only about 35 % of the moon was imaged in enough detail to create a usable map.

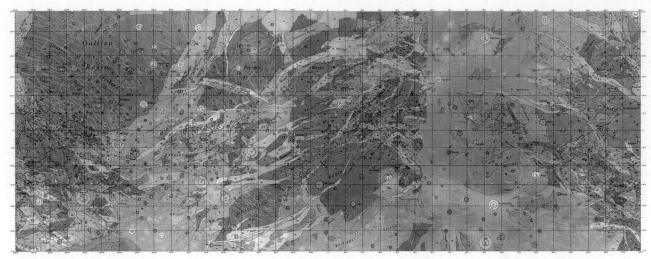

Fig. 10.5 Geologic map of Ganymede (Courtesy Collins, G. C., et al., U. S. Geological Survey)

Ideally, actual surface landing sites will require advanced study. When searching for ideal landing sites, spaceflight planners look for areas that will satisfy multiple scientific goals, while affording the crew a safe place to set down. Landing on most of the outer moons will be similar to the Apollo landings on Earth's Moon, a powered descent in a vacuum using some type of rocket engine.

For Jupiter's Galilean satellites, another consideration comes into play in landing site selection – radiation. "There isn't any place where you don't get some radiation," says Ames Research Center's Jeff Moore. "It's just a question of where do you get less radiation."

The worst case is the moon buried deepest in Jupiter's deadly magnetosphere, Io. "Io is such a high radiation environment," says volcanologist Rosaly Lopes, "It's just about the most inhospitable place in the Solar System." Landing on the pizza-faced moon may be hazardous to one's health, too. Active volcanoes erupt without warning, and are violent enough to damage an incoming lander. Additionally, the surface of Io is probably unstable in many areas, as extensive lava flows blanket vast plains with thin veneers that may not support the weight of a spacecraft, allowing it to break through to molten magma underneath.

Still, there may be locations suitable for human visits, if not habitation. Lopes points to underground lava tubes, caves that would provide protection from Jupiter's blazing radiation.

On Io, you would get the fatal dose of radiation in minutes. But we expect that there would be lava tubes because of the type of lava flows we see. I think they are pahoehoe flows similar to Hawai'i. However, it's a really different place in which to set up anything. You would have to keep people underground and then you would build your modules with oxygen and so on. You would have to be very sure that you're not going to get another volcanic eruption filling in those lava tubes, and how are you going to predict that? Volcanoes on Io are active for long periods of time, decades as far as we know. Even if there was an empty tube, are you going to go to all the trouble and expense to set up something now when there might be lava coming down the tube in the future?

Fig. 10.6 Some locations on Io probably resemble Hawai'i's Thurston lava tube, and could provide underground protection to human researchers from Jupiter's fierce radiation environment (Photo by the author)

More benign counterparts of Io's lava tubes will likely be found on the ice moons that have been cryovolcanically active. Worlds like Ganymede, Enceladus, Miranda and Ariel may well have flows of ice that have drained, leaving sheltered underground chambers just waiting for future settlers. These naturally pre-fabricated icehouses will be more stable in the geologically quiet environments that remain today.

Io's volcanism may not be all bad news for future visitors, says Lopes. Where there is heat, there is energy. "You could get geothermal power from the volcanoes themselves, but it would be tricky. It's even tricky on Earth; you can't get it from just any volcano. I don't know how far down you'd have to drill, but we think Io has this layer of a magma ocean underneath the crust. There may be heat conducted through the crust. But the easiest thing would really be to drill on the side of a caldera."

Fig. 10.7 Radiation levels vary across the face of Jupiter's moon Europa. The most sheltered areas are north and south of the equator on the leading hemisphere (Image courtesy of Wes Patterson and Chris Paranicas, Johns Hopkins Applied Physics Laboratory)

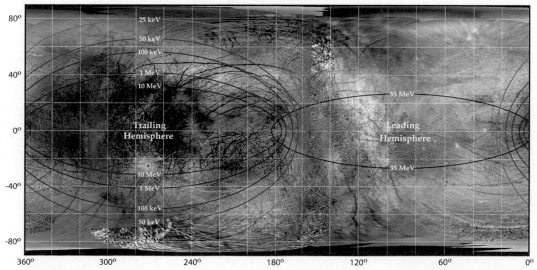

Less challenging from a radiation standpoint is the next moon out, Europa. While Io circles the king of worlds at a scant 421,700 km from the planet's center, Europa orbits Jupiter at a distance of 676,800 km, nearly twice the distance from Earth to the Moon. Nevertheless, the little ice satellite is still well within the Jovian magnetosphere. Radiation levels are about 540 REM per day, 40 REM more than a fatal dose. JPL's Robert Pappalardo explains:

> Most of that magnetosphere slams the back (of Europa). It has lots of effects. It does chemistry; it creates oxidants, and it sputters [wears away] the surface. Europa's surface is young enough that there hasn't been that much sputtering – that much erosion. In fact, that was one of the arguments for Europa being geologically young: "It can't be very old, because if the surface was very old, all these geologic features should have been sputtered away." You get the most erosion and radiation damage in the trailing hemisphere area, but then there is this weird line around the leading hemisphere where you get a band across the equator.

What does this mean for travelers? The best places to go for low radiation are at high latitudes, avoiding the trailing hemisphere.

For exploration, Europa has the added attraction of its subsurface ocean. Some 96 km of briny abyss lie beneath the frozen crust, and scientists would like to see what's down there. Probes with heating elements or drills have been tested in Antarctica's Lake Vostok and other sites, but getting through the ice while maintaining radio contact is a difficult chore, especially if that ice turns out to be 20 km thick. But eventually, technology will out, and it will be time for humans to go. Perhaps Europa's crust has thinner areas suitable for dropping a submersible through a borehole. Perhaps humans may even free-dive in pressure suits adapted to the alien environment.

Adventurers and researchers will find a similar situation at Saturn's Enceladus, where the subsurface ocean appears to be

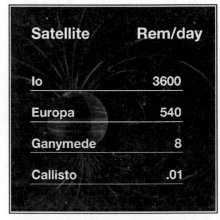

Satellite	Rem/day
Io	3600
Europa	540
Ganymede	8
Callisto	.01

Fig. 10.8

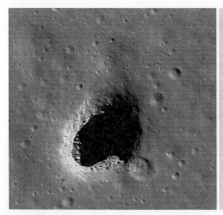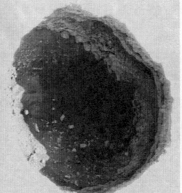

Fig. 10.9 Caves or lava tubes on other worlds may provide shelter from radiation. From left to right: Caves on Earth's Moon, Mars, and the Amirani lava plains of Io, thought to be cave-forming pahoehoe lava (Credits: Left: Image courtesy of NASA/LROC/ASU. Center: Image courtesy of NASA/JPL/University of Arizona. Right: Image courtesy of NASA/JPL-Caltech)

localized around the south pole. Engineers might erect a pressure dome first, in order to keep any water from boiling violently into the vacuum of space. Then, they would set to work drilling a tunnel through the ice crust. When the time was right, they would lower their spiffy submarine, complete with highly trained crew, into the depths of an alien sea.

The planet-moon Titan presents a unique environment with unique challenges for landing. Because of its dense, extended atmosphere, the descent from entry to the surface is nearly 3 h. On Mars, that familiar terror-filled landing sequence of rovers such as Opportunity and Curiosity was all over in 7 min. Titan is a far easier place to land. When the Huygens probe touched down on a parachute, it bounced lightly. Even if its last parachute had fouled, the probe probably would have slowed itself down enough to survive, simply because of the dense air and low gravity.

Landings on the mid-sized moons of the Solar System will be low-energy affairs. All moons in this size range have low densities and weak gravity. From a fuel standpoint, they will be easy targets for rest stops. But because of their bizarre low-gravity environments, we do not know quite what to expect on the surfaces. High-resolution images of Enceladus have shown "fluffy" areas surrounding the geysers, places where powdery snows might make landings a problem.

Even on massive Callisto, some slopes seem to be the consistency of talcum powder, which would mask landing operations (just ask the Lunar Module pilots). Fine dust may accrue in great quantities on many of the smaller moons. On the other end of the spectrum, Cassini has spotted boulders on the slopes of craters on Iapetus, Phoebe, Dione and other mid-sized moons. Landslides are evident on canyon walls on Uranus' Ariel, along with flows of material on valley floors. These may be rugged debris fields worth avoiding on final descent. Crisp edges of craters and faults on satellites such as Titania and Miranda seem to indicate a more "rocky" surface (despite the fact that those "rocks" are made of frozen water).

Large areas of some satellites may be more hard-packed than others. Cassini has discovered a pattern common to several mid-sized Saturnian moons – a wedge-shaped region of material around the equator that changes temperature more gradually than its surroundings. Scientists theorize that the Pac-Man thermal shape on the Saturnian moons occurs because high-energy electrons bombard low latitudes on the hemisphere of the moon that faces forward as it circles Saturn. The hail of electrons turns a powdery surface into hard-packed ice. As a result, the altered surface does not heat as rapidly in the sunshine or cool down as quickly at night as the rest of the surface. Both Mimas and Tethys have similar patterns, showing that high-energy electrons can dramatically alter the surface of an icy moon. Which terrain would make the best landing site remains to be seen. On the human scale, we do not yet have enough detail to tell what the landing hazards may be on the mid-sized moons.

Fig. 10.10 Surfaces of the small Saturnian moons vary in dust coverage and, consequently, in landing conditions. Note the flows of deep material on some of the surfaces of these small moons. Clockwise from top left: Helene, Methone, Calypso, Epimetheus (Cassini images courtesy of NASA/JPL/Space Science Institute)

NATURAL RESOURCES – SLIM PICKINGS?

Although astronauts may make use of Io's tidal heat and the Jovian electrical fields as resources, what else will be available to them? If humans are to remain for any time in the realm of the giants, they must find ways of living off the land.

At first blush, the worlds of gas and ice appear desolate and sterile. But it would not be the first time in history that explorers were faced with a shortage of in situ resources. A similarly daunting situation faced early settlers of Polynesia, says Jon Lomberg. "The Pacific – and the Hawai'i – that I know best is extremely resource-poor. There are no big animals. There are no metals; there is nothing that can be smelted. The islands are beautiful, but what you have to work with is extremely limited." In fact, building the voyaging canoe required every resource and every skill that the Polynesian culture had. The diaspora throughout the islands stood as the culmination of their society's technological evolution.

So it will be with the "voyaging canoes" of the outer system. Our transportation and the eventual infrastructure we build will challenge our best engineers and scientists. But there will be resources out there for the taking. Using those moons as bases, inhabitants will have resources close at hand: water for hydrogen and oxygen as sources of energy, and oxygen as a handy gas to breathe. Carbon dioxide provides a source of carbon to work with in manufacturing and for fuels. In the outer Solar System, says outer planets expert Heidi Hammel, "The Sun is so far away that I don't think you'll be using solar power. Who knows? You may be able to deploy massive things to collect sunlight. But more likely you'll be using radioisotopes – nuclear power – or some kind of clever chemical mechanism to make use of the *in situ* resources there." She adds: "Out there, it's so cold that those ices are not going to be soft. Ices are as hard as rock. You may be using them as your building materials. It's not that the pickings are slim; what's slim is our knowledge of those pickings."

As with Io's geothermal energy, there is a silver lining around the deadly features of Europa. Jupiter's radiation might be used as a power source. A study conducted at the European Space Agency, led by Dr. Claudio Bombardelli (now at Madrid Technical University), investigated the use of tethers to generate electricity within the Jovian magnetosphere. Their concept, which looked at power for orbiting spacecraft, was based upon the same principle as electric motors or generators – a wire moving through a magnetic field builds up an electrical current.

Bombardelli's team estimated that a spacecraft in polar orbit around Jupiter could generate a few kilowatts of power using two short (~3-km length) tethers. If those tethers were lengthened to 25 km, power generation could reach into the megawatt range. An outpost on one of the Galilean satellites would be farther from Jupiter than Bombardelli's case study, which assumed a closest distance (periapsis) of less than 1.3 Jupiter radii. Longer tethers would be required to generate power at the more distant moons, but these tethers could simply be laid along the surface of the moons like undersea cables. In this way, one of Jupiter's most deadly characteristics might also be an asset to future inhabitants.

On the moons of the giant worlds, water freezes to the consistency of rock. This, too, can be used to advantage. Just as Inuit peoples of the Arctic use ice to build igloos, so astronauts could use blocks from the icy surface to construct various structures. Under the conditions in the outer Solar System, ice will provide excellent masonry. The only caveat – it must be insulated from the interior heat. Humans like their rooms warm, and at a temperature of 21 °C, their walls and floors will begin to melt. "You have to isolate things, using insulation," says Lockheed Martin's Ben Clark. "It would be harder on Titan simply because the atmosphere's thick. But we do it on Earth. They insulate buildings from the cold surface in the Arctic and Antarctica." Another option would be inflatable habitats, says APL's Ralph Lorenz.

> You could do quite well with an inflatable dome on Titan. You'd want to anchor it down. There is enough wind to move particles around on the surface of Titan, as the sand dunes attest, and so a big inflatable would be a big area to drag on. So you'd stick your tent pegs in the side and you could have a fairly thin-walled dome. It's not going to have to deal with meteorite impacts or anything because the big

Fig. 10.11 *Arctic buildings like the Nullagvik Hotel in Kotzebue, Alaska, must be isolated from permafrost to keep them from melting into the ground. Note the spring-like radiators designed to shed excess heat from beneath the raised foundation, which is separated from direct contact with the ground by pilings (Photo by the author)*

atmosphere above you shields the surface from all that. What you would need to do is insulate it and stop the pressurized environment from melting through the ground. We don't know for sure what the surface composition of most of Titan is. In some places it may be water-ice. In most places it appears to be organic, so it's stuff that could soften or even melt at what we call room temperature. So you would need to insulate your Titan dome, but that's all you'd really need to do.

Inflatable habs will also find their uses in the vacuum environment of other moons at Jupiter, Saturn, Uranus and Neptune. Johnson Space Center has been conducting formal studies on habitats for the lunar surface for over a decade. With some added insulation, these technologies are directly applicable to the environments of the outer Solar System.

Aside from water, there may be other resources on the surfaces of the Galilean satellites. Callisto, for example, may have important resources for future settlers. Its dark component probably contains elements essential for life support, such as nitrogen, phosphorous, and carbon. Some silicon and common metals may also be present, but they are likely mixed up rather than in discrete deposits. The surfaces of Ganymede and Europa appear to be more pure water-ice, while rocky Io probably will offer visitors a host of sulfur compounds, silicates crystallized within lavas, and salts.

One satellite that seems to have more than its share of resources is Saturn's giant companion Titan, says JPL's Kevin Baines. "You've got all the hydrocarbons in the lakes, and you've got them falling from the sky as well. So the ground is probably impregnated with hydrocarbons anywhere you go there. If you want to use hydrocarbons for fuel, you don't need to go to those lakes; you can just dig it up anywhere. Titan would be the key place where you would set up your outpost. You've got all that water to break down into oxygen, hydrogen, carbon, methane, so you can use all the exotic materials to make rocket fuel pretty easily." As one aerospace analyst put it, "The fundamental elements of life and of plastics and industrial society are there."[1]

1. Robert Zubrin, personal conversation.

Fig. 10.12 The methane-powered Morpheus craft on a test flight at the Kennedy Space Center in Florida (Image courtesy of NASA)

METHANE MOVER

One of the most abundant resources in the outer Solar System is methane, and Titan is crawling with it. Recently, NASA's Johnson Space Center has joined forces with Armadillo Aerospace on the Morpheus planetary lander. The vertical takeoff and landing craft uses a combination of liquid oxygen and methane. Its current cargo capacity is a respectable 499 kg. Its Autonomous Landing Hazard Avoidance Technology (ALHAT) system enables the lander to perform pinpoint landings and to automatically avoid hazards, including steep slopes and boulders. The system incorporates a combination of laser altimeter, LIDAR and Doppler velocimeter.

As currently designed, the craft could carry rovers or propellant labs to the lunar surface, carry out rendezvous and propellant transfers in Earth orbit, and execute asteroid missions. But a similar craft could operate in the environment of Titan. There, natural resources could be processed as fuel for the vehicle. Surface ice could be broken down into hydrogen and oxygen, with the oxygen used for liquid propellant in combination with Titan's atmospheric methane to fuel the craft.

Titan is the only moon with an atmosphere of any substance. That atmosphere, like Earth's, is mostly nitrogen. The requirements for human survival on the surface are simple: an oxygen mask and a good fur coat. Second to Mars Titan is the easiest site to inhabit. Landings and launches require low energy, and the world is blanketed by useful materials lying around a relatively benign environment.

But that dense atmosphere is a two-edged sword when it comes time for departure. With that deep, dense atmosphere, astronauts will essentially need a slow rocket. Launching too fast in the lower atmosphere will waste precious energy and heat the exterior of the craft dangerously. Several scenarios have been discussed. One idea describes balloon-launched rockets. Another involves a flight design similar to Virgin Galactic's Spaceship One, in which an aircraft flies to high altitude, then deploys the rocket up to orbit.

Titan's environment will have other drawbacks. Humans living inside a habitat filled with oxygen will need to be careful of the naturally occurring materials outside their door, materials like butane, methane and propane. These would prove a dangerous mix to oxygen-loving residents. Ralph Lorenz says, "In principle, you could take your little trowel and bring in some dune sand from outside, pour it out on the table, and it probably would be flammable in an oxygen environment. That's a hazard that's dealt with in any coal mine. You would likely have to have brushes or a little anteroom where you would hose things down, but none of these things are terribly toxic in the short term." In short, none of Titan's challenges are show-stoppers for human exploration.

The poster child for futuristic space resources is helium-3. Helium-3 (^{3}He) is a light isotope of helium. The only place on Earth where it can be found is in the laboratory, but it is there for the taking on our nearby Moon, and it enriches the planets of the outer Solar System. Some believe that ^{3}He may provide all the power needs of our future, in the form of fuel for nuclear fusion.

Up to now, our nuclear energy has come from nuclear fission. In nuclear fission, the nucleus of atoms is split into smaller particles, releasing prodigious amounts of energy in the process. But a dangerous byproduct of fission is radioactive waste, piles of it. As we have seen at Russia's Chernobyl facility or the Fukashima plant in Japan, fission reactions can get away from the engineers. If a series of breakdowns occurs due to software or hardware problems, or if a natural disaster overwhelms the nuclear facility's fail-safes, disaster can happen.

Nuclear fission is constantly being held back because of its runaway issues, but fusion has the opposite problem. Left by itself, it shuts down rather than running away. This makes fusion intrinsically safer. It also promises to yield far more energy for outer planet spacecraft powered by nuclear propulsion.

Unlike a fission reaction, fusion forces together the atoms' nuclei so powerfully that their natural boundaries break down. To make this happen, the nuclei need to move at high speeds. When nuclear fuel is heated to a plasma, at temperatures of tens of millions of degrees, nuclei are freed to begin fusion. The only place in nature where fusion occurs is in the hearts of stars. There, high temperature atomic "soup" is held together by gravity. But in a fusion reactor, we must figure out how to hold things together at very high temperatures. Experiments are now taking place using strong magnetic fields, but so far, the process has been triggered for only milliseconds.

Current fusion attempts use deuterium, a heavy version of hydrogen. But deuterium has one serious drawback. Half of its energy leaks out through the magnetic field. That's where ^{3}He comes in. When combined with deuterium, the lost energy is something like 2 %, not 50 %. The good news? The outer Solar System is soaked in it. Saturn is the easiest of the giant planets from which to get ^{3}He. Its surface gravity is about the same as Earth, but it also has a very high rotational velocity, so that an atmospheric craft could actually reach orbit within known engineering concepts (with its massive gravity, Jupiter is beyond current capabilities). ^{3}He miners could descend into the upper atmosphere of Saturn, get their precious cargo and return to low orbit. Nuclear energy could heat the atmosphere in the jet engines used during mining operations. Hydrogen could also be liquefied while the craft flew through the atmosphere. That liquid [3]He would then be used as nuclear rocket fuel to get to orbit and to power other craft.

If fusion reactions can be sustained, they offer clean energy, essentially no radioactive waste, and a bright future! The problem, of course, is getting that fuel from the Moon or the outer planets to Earth.

The giant planets are much farther away. What would be our motivation for using them over the nearby Moon? The gravity of the gas and ice giants has trapped [3]He from the earliest days of the Solar System's formation. Lunar [3]He would satisfy the needs of Earth's population for several 1,000 years. But the [3]He available in the upper atmospheres of the giants would cover terrestrial demands for several billion years.[2] [3]He is of great interest in that all of its forms are "charged," which means that its exhaust can be directed with magnetic fields. Of any currently known physical system, says Robert Zubrin, "The Deuterium/helium-3 offers the greatest possibility of a rocket. It could actually get an exhaust velocity of on the order of 7 % of the speed of light. That is why fusion in general, and [3]He in particular, is of tremendous interest. It's not just an additional kind of power; it's a new kind of power."

TRAVELING IN STYLE

On most of the ice moons, the surface consistency may be similar to Earth's Moon, and the vacuum environment may call for vehicle designs similar to the Apollo lunar rovers with pressurized cabins added. Surface travel on Titan may require varied vehicles, just as it does on Earth. We have seen only one location from the surface of Titan, at the Huygens landing site. The texture of the ground there was apparently damp sand, with the appearance of a cobble-filled streambed. On Earth, many vehicles have been designed to handle such terrain. Large-wheeled or tracked rovers might be required if the methane-soaked sand is sticky. Titan's sand dunes bear a strong resemblance to the Arabian or Namib sand

2. *Our Cosmic Future* by Nikos Prantzos, Cambridge University Press 2000.

Fig. 10.13 An assortment of rovers are under development for operation in a vacuum. Rovers adapted to work on the icy moons may share some characteristics with these. Left: The Apollo lunar rovers were open to the space environment, and in the case of Apollo 17 traveled up to 36 km. (Image courtesy of NASA/JSC.) Center: The advanced Chariot rover (left) under study at NASA's Johnson Spaceflight Center can be adapted with a pressurized cabin (right). (Center left photo by author, center right by Reegan Geeseman, NASA.). Far right: An ESA student concept of a rover using walking legs for mobility (Image courtesy of ESA)

dunes on Earth, so some experts imagine desert-worthy land cruisers for those great sand seas. Dune travel might be challenging, as sand dunes are, literally, slippery slopes.

Titan's environment has much to offer future visitors in terms of travel, Ben Clark suggests. "One of the great things about Titan is that with the low gravity and dense atmosphere, balloons and even airplanes are simpler than anywhere else." Ralph Lorenz adds, "Titan affords lots of other opportunities for mobility on much larger scales. You can have a hot air balloon driven by the waste heat from a radioisotope generator, you can have a light gas balloon, or you can have an airship. You can fly heavier-than-air with an airplane. Helicopters would work well. Hovercraft have even been speculated." He adds: "On the seas, you can imagine vehicles floating. Hovercraft could go from one environment to another. You could even mention submarines. We know that one of the seas is at least 160 m deep, so there is the prospect for exploring the sea floor there with some sort of submersible. Almost any vehicle you can imagine makes sense at Titan. There are an awful lot of exciting possibilities."

TOURISM

Once we have learned to use the resources available among the giants, and once we have developed a network of infrastructure from the terrestrial planets to the outer system, people will begin to travel for reasons beyond science and exploration. They will travel for leisure and entertainment. Engineers and entrepreneurs will establish nodes at the locations best suited for navigation and richest in resources. In the Jovian system, these might be found at Ganymede and Callisto, while at Saturn, Titan's rich environment may be a strategic player. These three candidate moons have surface resources, but they also have a strong enough gravity well to be helpful in travel into and out of their immense planetary systems.

JPL's Kevin Baines envisions future tourism in the Saturnian system as being much like travel in the Caribbean or the Greek Islands today. "It would be a great tourist destination. It's like going to the Greek islands on a cruise ship. Each Greek isle is unique. It's the same with the moons of Saturn." Baines imagines day trips from the environs of Titan – perhaps orbiting transit stations – to the yin and yang landscapes of Iapetus, the battered highlands of Mimas or Tethys, and the geysers of Enceladus. "You would have cruise ships docking at different moon ports. Of course you would spend a day just hovering over the rings. You don't want to go through the rings, but [because of orbital mechanics] the little ship wants to go through them, so you'd have to keep firing your rockets to stay up."

Fig. 10.14 Future tours of the Saturnian system may resemble today's cruises among the Greek isles (Image © Sharon Malion. Used with permission)

Ben Clarke envisions such a satellite tour as taking place at breakneck speed. "To actually physically go to them, that's a lot of propellant. They're all in the equatorial plane, but you have to go and then come back again. You could do a multi-moon tour, like Cassini. You'd pass by fast, but I know people who like to watch Formula One, and it's kind of the same thing."

For future visitors to the giants, cruising the Saturn system will be no more bizarre than the idea of cruising the Caribbean would have seemed to medieval Europe. With some specific technological advancements, it will happen naturally.

There's even more to see out there. The diminutive Uranian satellites have some of the most spectacular geology in the Solar System. Mountain climbers will thrill to the sights from canyon walls, illuminated by the sapphire glow of Uranus. From Ariel, Uranus would appear as far across as 32 full Moons in Earth's sky. The 500-m-high volcanic flows that wander across the valley floors below, where cryolavas of liquid water froze in place eons ago, will attract visitors. Neighboring Miranda has an even more dramatic cliff. Verona Rupes is ten times the depth of North America's Grand Canyon, at roughly 20 km deep. Base jumping thrill seekers might enjoy the 12-min freefall to the bottom of the precipice, where they could cushion their landing with a rocket pack or airbags. Neptune's celebrity moon Triton provides a spectacular view of the cerulean ice giant.

Unlike the big moons in the Jupiter or Saturn system, which travel sedately around the equator with an unvarying equatorial view, Triton is blessed with a roller-coaster orbit. Its inclined path takes it high above the north pole, then brings it swooping down through the equator, under the south pole and back up again. Triton experiences this amazing sky show every 6 days. The view of Neptune changes dramatically as Triton's day progresses. Below that entertaining sky lies one of the most alien landscapes ever witnessed by human eyes. Jumbled cantaloupe terrain gives way to glistening ice plains peppered by impact craters, volcanic calderas, and cryolava flows. Beyond spreads the pink nitrogen polar ices, where Triton's unique geysers rise some 8 km overhead. At the horizon, travelers will see haze layers, but the sky will undoubtedly darken to a deep purple or black overhead in the rarified air.

ICE GIANT MISSION STUDIES

Many in the worldwide planetary science community view the ice giants as a priority in near-future robotic exploration. This is evidenced by the interest shown on the part of the world's space agencies. Studies have included flybys, minimal orbiters with probes, a set of simple orbiters, and Cassini-class missions to the ice giants, perhaps with Triton landers. Below is a selection of proposed missions. Note that most involve Uranus, as it is far easier to get to than distant Neptune.

- Uranus Pathfinder. ESA's medium-class orbiter, the craft would have been launched in 2020, perhaps with an atmospheric probe. Such a probe could measure the abundances of specific noble gases, telling scientists where Uranus formed initially and providing insight into details of scenarios like the Nice model. The proposal made it far along in ESA's selection process, but ultimately was turned down.
- NASA flagship mission to Uranus would cost over $1 billion (US). It would carry a large suite (~10) of instruments similar in scope to those on the Cassini Saturn orbiter. This mission would probably not be launched for two decades.
- NASA new frontiers-class mission to Uranus would cost as little as half what a flagship mission would cost. The spacecraft would be able to carry approximately five instruments. One advantage of this scale of mission is that it can be launched within a decade.
- NASA Discovery-class mission to Uranus would cost less than half a billion dollars but have a tightly focused mission consisting of essentially one instrument.
- Solar Electric Propulsion Orbiter/Probe. This mission would cruise to Uranus over a 13-year period with one gravity assist swing by from Earth. At Uranus, the orbiter would deploy a probe before a conventional chemical rocket slows it into orbit. The probe would enter the atmosphere at 22 km/s and slow by atmospheric drag. Once the probe is subsonic, it would deploy a parachute and transmit data from an altitude with a pressure of 0.1 bar (one tenth that of Earth at sea level) to 5 bars. Its cruise would last for about an hour before contact with the orbiter is lost.
- Neptune/Triton/Kuiper Belt mission. Four missions were studied by a U. S. group: (1) A single event flyby of Neptune that would then make its way on to a Kuiper Belt object; (2) A close-in Neptune orbiter that does not involve close encounters with Triton; (3) A 1-year mission focused on Triton itself; (4) A 2-year Triton-focused mission.
- ODINUS (Origins, Dynamics and Interiors of Neptunian and Uranian Systems). ESA twin spacecraft that would compare Uranus and Neptune. Each craft would settle into an eccentric, high orbit around their respective planets. Using an ion drive, they would spiral inward, studying the satellite systems and then the planets up close.

A NEW WILDERNESS

History has shown a progression from exploration to exploitation, and the latter has often been destructive to the natural beauty of our surroundings. As the outer Solar System becomes part of the human realm, how will we view it? How will we treat it? Wilderness has meant different things to different cultures. The concept of frontier, of a place for expansion and colonization and conquest, drove early European settlers in the Americas. The Western "wilderness" was seen, by many, as a source of natural resources or a playground rather than a place of natural beauty. Some even cultivated an adversarial attitude, seeing humankind's role as forging order out of the natural chaos. But the wilderness was also seen as a place of sanctuary, of mental and spiritual renewal, a realm that had aesthetic worth and intrinsic value.

This dichotomy of mindset was evident in the New World as European civilization took hold on the eastern seaboard. While scoping out practical resources and trade routes, early westward expeditions regularly brought artists along to capture the feel of the new lands and discoveries. The paintings of Albert Bierstadt and Thomas Moran helped to convince the U. S. Congress to found the first two national parks at Yellowstone and Yosemite. Frederick Catherwood documented the discoveries of Maya, Aztec and Incan ruins by John Lloyd Stevens. Frederick Church mounted expeditions to Antarctica, South America and other environs to create some of the most beautiful natural science paintings in history. Society saw a shift from a conqueror or colonial mentality to that of guardian.

Fig. 10.15 The painting View from Mount Holyoke, Northampton, Massachusetts, after a Thunderstorm – The Oxbow, by Thomas Cole, 1836. Note the small figure at lower center, and the clear cutting on the far slopes. (Metropolitan Museum of Art, Public domain)

Perhaps the clearest embodiment of this shifting attitude is seen in the Hudson River School of American painting from the nineteenth century. Hudson River artists depicted themes of exploration, discovery and the establishment of new frontiers, the very themes that will concern explorers of the giant worlds. Its members often portrayed small figures dwarfed by pastoral, idealized vistas, peacefully enjoying nature. At the time of the Hudson River art movement, the wilderness was rapidly vanishing from the Hudson Valley due to agriculture and urbanization. Ironically, the process occurred at the same time that local society was learning to cherish the stunning landscapes of the region.

Humankind stands at the opening of a vast new frontier. Some suggest that it is time to strategize how best to relate to it. "We have tried with mixed success to protect other vast wilderness areas, especially the polar regions and the seas," say authors Paul F. Uhlir and William P. Bishop. "The opportunity to protect the space wilderness is before us in the next few decades. We can and should set up the framework now."[3]

Do we twenty-first century humans see ourselves as conquerors of nature, using the environment around us merely to serve needs of humanity? (It is difficult not to think of the wilderness as something to be conquered when it consists of deadly vacuum and extreme temperatures!) Or do we see ourselves as an integral part of nature; do we perceive the wilderness as having intrinsic value? Do we stand somewhere in between?

As our technology progresses to the point at which we are capable of settling the outer Solar System, we must be aware of its power. Professor Holmes Rolston III, widely recognized as the father of environmental ethics, suggests six guidelines for future decisions on planetary exploration and settlement.[4]

1. Preserve and value natural sites that have earned proper names. Some proper names, he points out, are labels for the convenience of navigation, geology, etc. (for example, the Four Corners area or the Galileo Regio). Others are given names fairly arbitrarily. But some formations or sites warrant names because of their universally recognized significance, and it is these that should be preserved, he suggests.

2. Preserve and value exotic, extreme sites. Other worlds will express natural forms in ways that cannot exist on Earth or any other known world. These unique sites should be protected. Rolston explains, "Just as humans value diversity on Earth, humans should value diversity in the Solar System, all part of the robust richness of nature ... That a formative event in nature is rare is, *prima facie*, reason for its preservation. At such places, humans can learn something about the nature of things, the nature in things."

3. Preserve and value places of historical significance. For example, when humans return to the Moon, begin to homestead Mars or venture to the plains of Venus, what will become of the Apollo, Viking and Venera landing sites?[5] Some suggest that technology is now to the point where the

3. Paul F. Uhlir and William P. Bishop, "Wilderness in Space," from *Spaceship Earth: Environmental Ethics and the Solar System*, edited by Eugene C. Hargrove (Sierra Club Books, 1986).

4. *Spaceship Earth: Environmental Ethics and the Solar System* (Sierra Club Books, 1986).

5. In fact, a historic landing site has already been visited: the crew of *Apollo 12* landed within walking distance of the *Surveyor 3* lander, which had set down in the Ocean of Storms 31 months earlier. They removed the TV camera, soil sample scoop, and some aluminum tubing, returning the hardware to Earth for analysis.

Apollo lunar landing sites are already in danger of compromise. Google Lunar XPrize rules state: "In order to win this money, a private company must land safely on the surface of the Moon, travel 500 meters above, below, or on the Lunar surface, and send back two 'Mooncasts' to Earth. Teams may also compete for Bonus Prizes such as exploring lunar artifacts…" It was the last phrase that alarmed archaeologists and historians. Later, NASA and Google officials announced that "The race will abide by guidelines NASA has established to protect historic and scientifically important sites on the Moon." But herein lies another problem: NASA's guidelines regarding Apollo landing sites are Americentric. Their language deals with the sites in similar fashion to U. S. Park Service guidelines for American national parks. If we are to be successful in preserving significant historical sites, these sites must be recognized internationally.

4. Preserve and value places of active and potential creativity. Europa, Enceladus and Mars are all candidates for active biomes today. Until we know more about the nature of life and its possibility in these areas, Rolston suggests that they be considered off-limits for human exploration.

5. Preserve and value places of aesthetic importance. It is possible that many of the formations we have seen on the icy moons are fragile to the point that human traffic will seriously degrade or destroy them.

6. Preserve and value places of transformative importance. Rolston puts it this way: "Humans ought to preserve those places that radically transform perspective. Just as it was a good thing for medieval Europe to be dislodged from its insularity, challenged by the Enlightenment and the Scientific Revolution, it will be a good thing for Earthlings to be unleashed from the Earth-givens … Those who cannot be seriously confounded by nature have not yet seriously confronted it."

If our propulsion, medicine and engineering evolve as quickly over the next century as they have in the past one, and if the geopolitical landscape allows, explorers will be in the outer Solar System in the next century. At that point, humankind will be "seriously confounded" by its nature as we confront the worlds of gas and ice. Henry David Thoreau said: "We need the tonic of wildness…. At the same time that we are earnest to explore and learn all things, we require that all things be mysterious and unexplorable, that land and sea be indefinitely wild, unsurveyed and unfathomed by us…" But how will this wilderness affect us culturally and psychologically?

CULTURE AND CREATIVITY

Thomas Cole and Frederick Church will have their creative counterparts among those who will inhabit the realm of the giants. Wherever humankind goes, the species seems to leave a creative imprint, whether in petroglyphs, pyramids, monoliths or henges. The environments of the outer planets and their moons will inspire new ways of expressing the creativity that seems to be such an intrinsic part of being human. JPL's Robert Pappalardo points to Europa as one such example.

> Jupiter is going to be fixed in the sky, and you'd want to have your lawn chair facing that way to watch the changing show. Europa is librating slightly, so Jupiter will nod up and down a bit, sort of shifting around, because Europa itself is nodding around [in its orbit]. Jupiter won't rise and set, but it will move. You could probably tell what time it was by the position of Jupiter, something like a sundial. At the same time, you'll see the phases of Jupiter go by. You'll have a nice bright Jupiter at night on your horizon. During the day it would be mostly dark, but you might see the rings in the sunlight.

Perhaps one day, a handful of human artisans will set up something akin to Robert Pappalardo's Europa sundial, a henge to keep track of the local time by the position of Jupiter on the horizon. With advanced computers nestled inside space helmets, it would not be necessary, but it would be wonderful.

Jon Lomberg cautions that – if the past is any indication – it will be difficult to project what forms human creativity will take in the future.

> If you were in the Renaissance and trying to imagine what painting would be like in the 21st century; if you were really imaginative you might envision different kinds of paints: water-soluble paints and all the great painting materials we have. But would you think of computer graphics? I don't think your mind would ever go

Fig. 10.16 Stonehenge, Europa style. Because of Europa's libration, stones could be erected in such a way as to mark the time of day by the position of Jupiter in the sky (Painting © Michael Carroll)

in that direction because there are just too many jumps. I think there will still be people who are knitting sweaters and painting watercolors and preserving all of our techniques. But that will be considered by the artists of the time as so retro. The thing about art is it doesn't become obsolete the way technology does. Old technology is of a historical interest, but new technology works better. When you look at those paintings of animals in the caves you can put them up against any animals painted by any artists since and they'll hold their own. The old mediums and the old interactions, I'd like to think, won't be abandoned. There will always be people who'd rather get their hands into clay than use a 3D printer.

THE GIANTS AWAIT

While studying a painting of the rings of Saturn, Cassini's Carolyn Porco remarked, "Sometimes I get discouraged, and then I look at a picture like that and I think, 'Oh my God, I just want to be there!'" Space exploration is more than the return of data. It is inspiration. It is stirring. Icy moons expert Tilmann Denk sees a profound purpose in the work he does. "In my view, the exploration of the Solar System is a cultural achievement, and any attempt to couple the 'dry findings of knowledge' with the senses of regular human beings is not just highly appreciated, but the deep reason for our work."

Many share Tilmann Denk's perspective. There seems something significant about the exploration of new frontiers, about the pushing of technology and people to their limits. Thousands of people have sacrificed much so that humankind could visit the giant worlds vicariously. For many, space exploration is the most positive of human endeavors.

The people at the tip of the spear, those who would dare to look a giant in the face, are dreaming and planning and hard at work all over Earth. It's a long way to the outer realm of the giants, but one day, advances in life sciences, propulsion, our knowledge of planetary geology, and enabling technologies for long-term habitation in cryogenic environments will see humankind's eventual arrival in the outer worlds. A new frontier awaits.

In the ancient book of *Numbers*, the narrative says, "And there we saw the giants, the sons of Anak, which come of the giants: and we were in our own sight as grasshoppers, and so we were in their sight."[6]

When we stand before giants, we are humbled. The immense power, majestic beauty, terrifying cyclones, furious radiation, elegant aurorae, and vast moon systems of the outer giant planets dwarf all that is familiar to us in our terrestrial realm. And yet, we gain motivation from them as well, inspiration to explore, to push our technologies and our influences and our frontiers as a species. In turn, the technologies inspired by those giants enhance our quality of life on Earth, and the wilderness of the worlds of methane, hydrogen and ice inform our culture, our society, our arts and our perspectives. They can do no less than enrich our lives, and they will continue to do so — ever more deeply — as we venture out to live among the giants.

6. Numbers 13:33, King James Version of the Holy Bible.

Index

M. Carroll, *Living Among Giants: Exploring and Settling the Outer Solar System*,
DOI 10.1007/978-3-319-10674-8, © Springer International Publishing Switzerland 2015

Printed by Books on Demand, Germany